COLEBROOKE'S
TRANSLATION OF THE

LILAVATI

COLEBROOKE'S
TRANSLATION OF THE

LILAVATI

WITH NOTES BY

H. C. BANERJI

SECOND EDITION

GYAN

Published by

Gyan Publishing House
5, Ansari Road
Daryaganj, New Delhi-110002
Phone: 011-47034999, 9811692060
E-mail: books@gyanbooks.com

Distribution Network
gyanbooks.com
India, USA, Canada, UK, Australia, France

ISBN : 978-81-212-3727-7 (PB)
First Published, 1927

2nd Impression 2020

Printed at: Gyan Press, Delhi.

Lilavati
Author: H. C. Banerji

COLEBROOKE'S TRANSLATION

OF THE

LILÁVATI.

WITH NOTES

BY

HARAN CHANDRA BANERJI, MA., B.L.

Second Editio

CALCUTTA

The Book Company, Ltd.

1927

PREFACE.

THIS little book is an edition of Colebrooke's translation of the *Líláватí*, a standard work on Hindu mathematics, written by Bháskaráchárryya, a celebrated mathematician and astronomer who lived in the twelfth century of the Christian era.[1] The work forms the first part of a larger work of the author called the *Siddhánta-siromani*. This part is called by the author, *Pátíganita* or Arithmetic ; but this name has not been properly given. For, the work, besides dealing with subjects which lie within the province of Arithmetic, also treats of subjects which properly belong to Algebra and Geometry. It comprises the four simple rules, extraction of the square root and the cube root, vulgar fractions, Rule of Three, interest, alligation, problems producing simple and quadratic equations, arithmetcial and geometrical progressions, permutations and combinations, indeterminate equations of the first degree, several properties of triangles and quadrilaterals, areas of circles, volumes of spheres, cones and pyramids, solid content of excavations, and several

[1] This date is ascertained from the fact that Bháskara himself informs us in a passage of his *Siddhánta-siromani*, that he was born in the year 1036 of the *Saka* era, and that he completed his great work when he was 36 years old. This gives 1150 A.D., as the date of the completion of the *Siddhánta-siromani*. See the *Goládhyáya* of the *Siddhánta-siromani*, Wilkinson's translation, XIII, 58.

other matters. Some of the problems solved evince a great deal of progress in algebraical investigations. The author does not state the reasons for the various rules given by him. I have tried to supply the reasons as simply and shortly as they occurred to me ; but still in some cases neater and shorter demonstrations may possibly be given. The explanations given have been printed in small type and enclosed within square brackets. It is thus hoped that the present edition will prove useful and interesting not only to the scholar and the antiquarian, but also to the student of modern algebra.

In his foot-notes, Colebrooke has given translations of extracts from the leading commentaries on the *Lílávatí*. These are :—(1 The commentary of Gangádhara, written about 1420 A.D.: (2) that of Súryadása, called *Ganitámrita*, written in 1538 A.D., containing a clear interpretation of the text, with concise explanations of the rules : (3) that of Ganesa, called *Buddhivilásiní*, the best of all the commentaries, written in 1545 A.D., comprising a copious exposition of the text, with demonstrations of the rules : (4) the gloss of Ranganátha on the *Vásaná*, Bháskara's demonstratory annotations of the *Siddhánta-siromani*, written towards the beginning of the seventeenth century A.D. : (5) the *Manoranjana*, written by Rmá Krishna Deva, of uncertain date : and (6) the *Ganitakaumudí*, which has not been recovered, but is known from the quotations cited from it by Súryadása and Ranganátha. Some of the translated extracts contain expositions of the rules and of technical terms, and some contain demonstrations of the rules in a few cases. Of these demonstrations which are given chiefly by Ganesa and Súryadása, those which are satisfactory and instructive, have been retained in the present edition; whilst others which are obscure and unsatisfactory, have been omitted. For convenience of reference, the *Lílávatí* in Sanskrit is printed at the end, with divisions into chapters and sections corresponding to those made in the translation. No such divisions were made by Bháskara.

NARIKELDANGA, CALCUTTA,
December, 1892.

H. C. B.

PREFACE TO THE SECOND EDITION.

IN this edition, the book has been carefully revised and a few slight additions and alterations have been made in the notes.

NARIKELDANGA, CALCUTTA,
November, 1927.

H. C. B.

CONTENTS.

LÍLÁVATI.

CHAPTER I.

INTRODUCTION.

1. Having bowed to the deity, whose head is like an elephant's[1]; whose feet are adored by gods; who, when called to mind, relieves his votaries from embarrassment; and bestows happiness on his worshippers; I propound this easy process of computation,[2] delightful by its elegance,[3] perspicuous with words concise, soft and correct, and pleasing to the learned.

DEFINITIONS OF TECHNICAL TERMS.

(Money by tale.)

2. Twice ten cowry shells[4] are a *kákiní;* four of these are a *pana*; sixteen of which must be here considered as a *dramma;* and in like manner, a *nishka*, as consisting of sixteen of these.

---

[1] Ganesa, represented with an elephant's head and human body.

[2] *Patiganita*; *pati*, *paripati*, or *vyaktaganita*, arithmetic.

[3] *Lilavati*, delightful: an allusion to the title of the book. See notes on §§ 13 and 277.

[4] Cypræa moneta. Sans., *Varataka*, *kapardi*, Hindi, *Kauri*.

(*Weights.*)

3. A *gunja*[1] (or seed of *Abrus*) is reckoned equal to two barley-corns; a *valla*, to three *gunjas*; and eight of these are a *dharana* ; two of which make a *gadyánaka*. In like maner one *dhataka* is composed of fourteen *vallas*.

4. Half ten *gunjas* are called a *másha*, by such as are conversant with the use of the balance; a *karsha* contains sixteen of what are termed *máshas*; a *pala*, four *karshas*. A *karsha* of gold is named *suvarna*.

(*Measures.*)

5—6. Eight breadths of a barley-corn[2] are here a finger; four times six fingers, a cubit[3]; four cubits, a staff[4]; and a *krosa* contains two thousand of these; and a *yojana*, four *krosas*.

So a bambu pole consists of ten cubits; and a field (or plane figure) bounded by four sides, measuring twenty bambu poles, is a *nivartana*.[5]

---

1 A seed of Abrus precatorius, black or red; the one called *Krishnala*, the other *raktika*, *ratti*, or *rattika*.

2 Eight barley-corns (*yasa*) by breadth, or three grains of rice by length are equal to one finger (*angula*).—Gan.

3 According to Ganesa, the cubit (*hasta*) means the practical cubit employed by artisans and called *gaj*. It is longer than the ordinary cubit of 18 inches.

4 *Danda*, a staff: directed to be cut nearly of man's height. (Manu, II. 46.)

5 A superficial measure containing 400 square poles.—Sur.

7. A cube,[1] which in length, breadth and thickness measures a cubit, is termed a solid cubit : and, in the meting of corn and the like, a measure, which contains a solid cubit, is a *khárí* of *Magadha*[2] as it is denominated in science.

8. A *drona* is the sixteenth part of a *khárí;* an *ádhaka* is a quarter of a *dorna ;* a *prastha* is a fourth part of an *ádhaka ;* and a *kudaba* is by the ancients termed a quarter of a *prastha.*[3]

The rest of the axioms, relative to time[4] and so forth, are familiarly known.

---

[1] *Dwadasasra*, lit. dodecagon, but meaning a parallelopiped : the term *asra*, corner or anglo, being here applied to the edge or line of incidence of two planes.

[2] The country situated on the *Sonebhadra* river.—Gan. It is South Behar.

[3] Another stanza occurs here in one copy of the text. It is explained in the *Manoranjana*, and by Gangadhara, but not by Ganesa and Suryadasa. It is therefore to be rejected as spurious and interpolated. It is as follows :—"The *sera* is here reckoned at twice seven *tankas*, each equal to three-fourths of a *gadyanaka :* and a *mana*, at forty *seras*. The name is in use among the *Turushkas*, for a weight of corn and like articles." See notes on §§ 97 and 233.

[4] The author has himself explained the measures of time in his *Siddhantasiromani*. [See the *Goladhyaya*, Wilkinson's translation, IV. 5—12.—Ed.]

CHAPTER II.

SECTION I.

INVOCATION.[1]

9. Salutation to Ganesa, resplendent as a blue and spotless lotus ; and delighting in the tremulous motion of the dark serpent, which is perpetually twining within his throat.

NUMERATION.

10—11. Names of the places of figures have been assigned for practical use by ancient writers,[2] increasing regularly[3] in decuple proportion : namely, unit, ten, hundred, thousand, myriad, hundred thousands, million, ten millions, hundred millions, thousand millions, ten thousand millions, hundred thousand millions, billion, ten billions, hundred

[1] A reason of this second introductory stanza is, that the foregoing definitions of terms are not properly a part of the treatise itself ; none such having been promised by Arya Bhatta and other ancient authors in their treatises on arithmetic.—Gan. and *Mano.*

[2] According to the Hindus numeration is of divine origin ; the invention of nine figures (*anka*), with the device of places to make them suffice for all numbers, being ascribed to the beneficent Creator of the Universe, in Bhaskara's *Vasana* and its gloss ; and in Krishna's Commentary on the *Vijaganita.* Here nine figures are specified ; the place, when none belongs to it, being shown by a blank (*sunya*), which, to obviate mistake, is denoted by a dot or small circle.

[3] From the right, where the first and lowest number is placed, towards the left hand —Gan.

billions, thousand billions, ten thousand billions, hundred thousand billions.[1]

SECTION II.

EIGHT OPERATIONS[2] OF ARITHMETIC.

12. Rule of addition and subtraction[3] : half a stanza.

The sum of the figures according to their places is to be taken in the direct or inverse order[4] : or (in the case of subtraction) their difference.

[The rule as exemplified in the *Manoranjana* is more cumbrous than the ordinary rule.]

13. Example. Dear intelligent Lilavati,[5] if thou be skilled in addition and subtraction, tell me the sum of two, five, thirty-two, a hundred and ninety-

---

[1] Sans, *eka, dasa, sata, sahasra, ayuta, laksha, prayuta, koti, arbuda, abja* or *padma, kharva, nikharva, mahapadma, sanku, jaladhi or samudra, antya, madhya, parardha.*

A passage of the *Veda*, which is cited by Suryadasa, contains the places of figures :—"Be these the milch kine before me, one, ten, a hundred, a thousand, ten thousand, a hundred thousand, a million, ... Be these milch kine my guides in this world."

[2] *Parikarmashtaka*, eight operations, or modes of process ; logistics or algorism.

[3] *Sankalana, sankalita, misrana, yuti, yoga, &c.*, summation, addition, *Vyavakalana, vyavakalita, sodhana, patana, &c*, subtraction. *Antara*, difference, remainder.

[4] From the first on the right, towards the left ; or from the last on the left, towards the right.—Gang.

[5] Seemingly the name of a female to whom instruction is addressed. But the term is interpreted in some of the commentaries, consistently with its etymology, 'charming.'—See §§ 1 and 277.

three, eighteen, ten and a hundred, added together; and the remainder, when their sum is subtracted from ten thousand.

Statement : 2, 5, 32, 193, 18, 10, 100.

Result of the addition[1] : 360.

Statement for subtraction : 10000, 360.

Result of the subtraction : 9640.

14—15. Rule of multiplication[2] : two and a half stanzas.

Multiply the last[3] figure of the multiplicand by the multiplicator, and next the penult, and then the rest, by the same repeated. Or let the multiplicand be repeated under the several parts of the multiplicator, and be multiplied by those parts : and the products be added together. Or the multiplier being divided by any number which is an aliquot part of it, let the multiplicand be multiplied by that number, and then by the quotient, the result is the product. These are two methods of subdivision by form. Or multiply separately by the places of figures, and add the products together. Or multiply by the mul-

---

[1] Mode of working addition as shown in the *Manoranjana* :

Sum of the units, 2,5,2,3,8,0,0,	...	...	...	2	0
Sum of the tens, 3,9,1,1,0,	...	...	... 1	4	
Sum of the hundreds, 1,0,0,1,	...	...	... 2		
Sum of the sums ...	...	...	... 3	6	0

[2] *Gunona, abhyasa* ; also *hanana* and any term implying a tendency to destroy. It is denominated *pratyutpanna* by Brahmagupta and by Sridhara. *Gunya,* multiplicand. *Ganaka,* multiplicator. *Ghata,* product.

[3] The digit standing last towards the left.

tiplicator diminished or increased by a quantity arbitrarily assumed; adding or subtracting the product of the multiplicand taken into the assumed quantity.

[The author gives here six methods. The first method is the ordinary one, and includes the *tatstha* of the older authors, which is worked by repeating or moving the multiplier over or under every digit of the multiplicand, and which, according to Ganesa's explanation, proceeds obliquely, joining products along compartments. The second is tedious, following from the formula, $a(b+c)=ab+ac$. The third is multiplication by factors. The fourth is practically the same as the first, and the fifth the same as the second. The sixth follows from the formula $a(b-c)=ab-ac$.]

16. Example. Beautiful and dear Lilavati, whose eyes are like a fawn's! tell me numbers resulting from one hundred and thirty-five, taken into twelve, if thou be skilled in multiplication by whole or by parts, whether by subdivision of form or separation of digits.[1] Tell me, auspicious woman, the

[1] The adjoined scheme of the process of multiplication is exhibited in Ganesa's commentary.

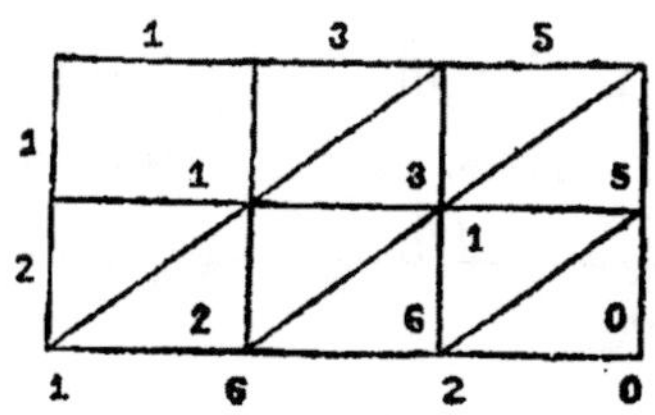

According to the *tatstha* method, the process will stand thus :—

$$\begin{array}{ccc} 12 & 12 & 12 \\ 1 & 3 & 5 \\ \hline 12 & & 60 \\ & 36 & \\ \hline & 1620 & \end{array} \quad \text{or} \quad \begin{array}{cc} 135 & 135 \\ 1 & 2 \\ \hline & 270 \\ & 135 \\ \hline & 1620 \end{array}$$

quotient of the product divided by the same multipiler.

Statement: Multiplicand 135. Multiplicator 12.

Product (multiplying the digits of the multiplicand successively by the multiplicator) 1620.

Or, subdividing the multiplicator into parts, as 8 and 4; and severally multiplying the multiplicand by them; adding the products together, the result is the same 1620.

Or, the multiplicator 12 being divided by 3, the quotient is 4; by which, and by 3, successively multiplying the multiplicand, the last product is the same, 1620.

Or, taking the digits as parts *viz.*, 1 and 2; the multiplicand being multiplied by them severally, and the products added together, accordiug to the places of figures, the result is the same 1620.

Or, the multiplicand being multiplied by the multiplicator less 2, *viz.*, 10, and added to twice the multiplicand, the result is the same, 1620.

Or, the multiplicand being multiplied by the multiplicator increased by 8, *viz.*, 20, and eight times the multiplicand being subtracted, the result is the same, 1620.

17. Rule of division[1]: one stanza. That number, by which the divisor being multiplied balances

---

[1] *Bhaga-hara*, *bhajana*, *harana*, *chhedana*, division. *Bhajya*, dividend. *Bhojaka*, *hara*, divisor. *Labdhi*, quotient.

the last digit of the dividend (and so on[1]), is the quotient in division: or, practicable, first abridge[2] both the divisor and the dividend by an equal number, and proceed to division.

Example. Statement of the number produced by multiplication in the foregoing example, and of its multiplicator, for a dividend, 1620, and a divisor, 12.

Quotient 135; the same with the original multiplicand[3].

Or both the dividend and the divisor, being reduced to least terms by the common measure 3, are 540 and 4; or by the common measure 4, they become 405 and 3. Dividing by the respective reduced divisors, the quotient is the same, 135.

[The first part of the rule is vague and incomplete, although it is practically the same as the ordinary rule, as will be evident from the foot-note 3, p. 9. The second part follows from the identity $ab \div ac = b \div c$.]

---

[1] Repeating the divisor for every digit, like the multiplier in multiplication.—Gang.

[2] *Apavartya*, abridging. See note on § 249.

[3] The process of long division is exhibited in the *Manoranjana*, thus: The highest places of the proposed dividend, 16, being divided by 12, the quotient is 1; and 4 over. Then 42 becomes the highest remaining number, which divided by 12 gives the quotient 3, to be placed in a line with the preceding quotient 1: thus 13. Remainder 60, which divided by 12 gives 5: and this being carried to the same line as before, the entire quotient is exhibited: *viz.*, 135.

18—19. Rule for the square[1] of a quantity : two stanzas.

The multiplication of two like numbers together is the square. The square of the last[2] digit is to be placed over it ; and the rest of the digits, doubled and multiplied by that last, to be placed above them respectively ; then repeating the number, except the last digit again (perform the like operation). Or twice the product of two parts, added to the sum of the squares of the parts, is the square (of the whole number.)[3] Or the product of the sum and difference of the number and an assumed quantity, added to the square of the assumed quantity, is the square.[4]

["The square of the last digit &c." The translation here is slightly incorrect. It should run thus :—"The square of the last digit, and the rest of the digits doubled and multiplied by that last, are to be placed one above the other (regard being had to the local values) ; then repeating, &c." It will then appear that the first two methods are really the same, and are based on the formula, $(a+b+c)^2=a^2+2a(b+c)+b^2+2bc+c^2$. The

[1] *Varga, kriti*, a square number.

[2] The process may begin with the first digit, as intimated by the author in § 24.

[3] The proposed quantity may be divided into three parts, instead of two; and the products of the first and second, first and third, and second and third, being added together and doubled, and added to the sum of the squares of the parts, the total is the square sought.—Gan.

[4] Another method is hinted in the author's note on this passage; consisting in adding together the product of the proposed quantity by any assumed one, and its product by the proposed less the assumed one.—Rang. [This follows from the identity, $ab+a(a-b)=a^2$.—Ed.]

working of the first method is this :—Suppose we have to find the square of 297.

Then

7^2	= 49	or,	2^2		4
$7\times2\times29$	= 406		$2\times2\times97$	=	388
9^2	= 81		9^2	=	81
$9\times2\times2$	= 36		$9\times2\times7$	=	126
2^2	= 4		7^2	=	49
$\therefore$ $(297)^2$	= 88209 ;		$\because$ $(297)^2$	=	88209

The ciphers are omitted for simplicity. The third method follows from the identity, $(a+b)(a-b)+b^2=a^2$, a being the proposed and b, the assumed quantity.]

20. Example. Tell me dear woman, the squares of nine, of fourteen, of three hundred less three, and of ten thousand and five, if thou know the method of computing the square.

Statement : 9, 14, 297, 10005.

Proceeding as directed, the squares are found : 81, 196, 88209, 100100025. Or, put 4 and 5, parts of 9. Their product doubled 40, added to the sum of their squares 41, makes 81. So, taking 10 and 4, parts of 14, their product 40, being doubled, is 80 ; which, added to 116, the sum of the squares 100 and 16, makes the entire square, 196.

Or putting 6 and 8, their product 48, doubled, is 96 ; which, added to the sum of the squares 36 and 64, *viz.*, 100, makes the same 196.

Again, 297, diminished by 3, is 294 ; and, in another place increased by the same, is 300. The product of these is 88200 ; to which adding the square of 3, *viz.*, 9, the sum is as before the square, 88209.

21. Rule for the square, root[1] : one stanza.

Having deducted from the last of the odd digits[2] the square number, double its root; and by that dividing the subsequent even digit, and subtracting the square of the quotient from the next uneven place, note in a line (with the preceding double number) the double of the quotient. Divide by the (number as noted in a) line the next even place, and deduct the square of the quotient from the following uneven one, and note the double of the quotient in the line. Repeat the process (until the digits be exhausted). Half the (number noted in the) line is the root.

[The rule is practically the same as the ordinary one for the extraction of the square root. Only it is a little more cumbrous, as will appear from the two processes placed side by side :—

```
          1_1_1
          8 8 2 0 9 (4                          . . .
          4          18                         8 8 2 0 9 (297
          -----      ---                        4
          482        58                   49 | 482
 4×9 =    36          14                     | 441
          -----      ----                    | ----
          122        594                     | 4109
 9² =      81        ½ of 594            587 | 4109
          -----      =297,                     ----
          4109       the root required.
 7×58 =   406
          -----
           49
 7² =      49
```

[1] *Varga múla*, root of the square; *múla, pada*, are synonyms of root.

[2] Every uneven place is to be marked by a vertical line, and the intermediate even digits by a horizontal line. But, if the last place be even, it is joined with the contiguous odd digit. Example, $\overset{1\_1\_1}{88209}$.

Thus we see that instead of directing the subtraction of 9×49 at once from 482, the rule directs first the subtraction of 9×40, and then from the remainder the subtraction of 9×9 or 9^2. And similarly for the next step. The process shown on the left-hand side is the same as that explained in the *Manoranjana.*]

22. Example. Tell me dear woman, the root of four, and of nine, and those of the squares before found, if thy knowledge extend to this calculation.

Statement : 4, 9; 81, 196, 88209, 100100025. The roots are 2, 3, 9, 14, 297, 10005.

23—25. Rule for the cube[1] : three stanzas.

The continued multiplication of three like quantities is a cube. The cube of the last (digit) is to be set down ; and next the square of the last multiplied by three times the first ; and then the square of the first taken into the last and tripled ; and lastly, the cube of the first : all these, added together according to their places, make the cube. The proposed quantity (consisting of more than two digits) is distributed into two portions, one of which is then taken for the last (and the other for the first) ; and in like manner repeatedly (if there be occasion.)[2] Or the same process may be begun from the first place of figures, either for finding the cube or

---

[1] *Ghana*, a cube ; lit., solid.

[2] The subdivision is continued until it comes to single digits. Ganesa confines it to the places of figures (*sthána-vibhága*), not allowing the portioning of the number (*rúpa-vibhága*), because the addition is to be made according to the places.

the square. Or three times the proposed number, multiplied by its two parts, added to the sum of the cubes of those parts, give the cube. Or the square root of the proposed number being cubed, that multiplied by itself, is the cube of the proposed square.[1]

[The different methods follow from the following formulæ :—

$$(a+b)^3 = a^3 + 3a^2b + 3ab^2 + b^3.$$

$$(a+b)^3 = a^3 + b^3 + 3ab(a+b).$$

$$\left\{(\sqrt{a^2})^3\right\}^2 = (a^2)^3.$$]

26. Example. Tell me, dear woman, the cube of nine, the cube of the cube of three, and the cube of the cube of five, and the cube roots of these cubes, if thy knowledge be great in the computation of cubes.

Statement : 9. 27, 125.

The cubes in the same order are, 729, 19683. 1953125.[2]

---

[1]This carries an allusion to the raising of quantities to higher powers than the cube. Ganesa specifies some of them. Thus the fourth power of a number is called *varga-varga* ; the sixth power, *varga-ghana* or *ghana-varga* : the eighth power, *varga-varga-varga* ; the ninth power, *ghana-ghana* ; the fifth power, *varga-ghana-ghâta;* and the seventh power, *varga-varga-ghana-ghâta.*

[2]The following process of finding the cube of 125 is given in the *Manoranjana.* The proposed number 125 is distributed into two parts 12 and and the first of these again into two parts 1 and 2:

The proposed number being 9, and its parts 4 and 5, 9 multiplied by them and by 3 is 540; which, added to the sum of the cubes 64 and 125, *viz.*, 189, makes the cube of 9 *viz.*, 729. The entire number being 27, its parts are 20 and 7; by which the number being successively multiplied, and then tripled, is 11340; and this added to the sum of the cubes of the parts, 8343, makes the cube 19683.

The proposed number being a square as 4, its root 2 cubed is 8. This taken into itself gives 64, the cube of 4. So 9 being proposed, its square root 3, cubed, is 27; the square of which, 729, is the cube of 9. In short, the square of the cube is the same as the cube of the square.

["In short, tne square, &c." The translation should be, "in short, the cube of a square number is the same as the square of the cube of the square root of the number." This follows from the third formula given in the preceding article.]

27—28. Rule for the cube root[1]: two stanzas.

Then, 1 cubed is	1
1, square of 1, tripled and multiplied by 2, is ...	6
4, square of 2, tripled and multiplied by 1, is ...	12
2 cubed is	8
	1728
Now, 12 cubed as above is	1728
144, square of 12, tripled and multiplied by 12, is ...	2160
25, square of 5, tripled and multiplied by 12, is ...	900
5 cubed is	125
Thus, 125 cubed is	1953125

[1] *Ghana-múla*, root of the cube.

The first (digit) is a cube's place; and the two next, uncubic; and again, the rest in like manner. From the last cubic place take the (nearest) cube, and set down its root apart. By thrice the square of that root divide the next (or uncubic) place of figures, and note the quotient in a line (with the quantity before found). Deduct its square taken into thrice the last (term), from the next (digit); and its cube from the succeeding one. Thus the line (in which the result is reserved)- is the root of the cube. The operation is repeated (as necessary).

Example. Statement of the foregoing cubes for extraction of the root: 729, 19683, 1953125.

The cube roots respectively are 9, 27, 125.

[The rule is more cumbrous than the ordinary one, as will appear below :—

1953125 (1, 2, 12, 5, 125

$1^3 = 1$

953125

$3 \times 1^2 = 3$

$3 \times 2 = 6$

353125

$3 \times 2^2 \times 1 = 12$

233125

$2^3 = 8$

225125

$3 \times 12^2 = 432$

$5 \times 432 = 2160$

9125

$5^2 \times 12 \times 3 = 900$

125

$5^2 = 125$

1953125 (125

1

$3 \times 10^2 = 300$ | 953

$3 \times 10 \times 2 = 60$

$2^2 = 4$

364 | 728

225125

$3 \times 120^2 = 43200$

$3 \times 120 \times 5 = 1800$

$5^2 = 25$

45025 | 225125

[The process shown on the left-hand side is the same as that explained in the *Manoranjana*. The ciphers are omitted for simplicity.]

SECTION III.

FRACTIONS.[1]

FOUR RULES FOR THE ASSIMILATION OR REDUCTION OF FRACTIONS TO A COMMON DENOMINATOR.[2]

29. Rule for the simple reduction of fractions[3]: one stanza. The numerator and denominator[4] being multiplied reciprocally by the denominators of the two quantities,[5] they are thus reduced to the same denominator. Or both numerator and denominator may be multiplied by the intelligent calculator into the reciprocal denominators abridgod by a common measure.

---

[1] *Bhinna*, a fraction ; lit., a dividod quantity, or ono obtained by division.—Gan. An incomplete quantity or non-integer (*apúrna*).—Gang.

[2] *Bhága-játi-chatushtaya, játi-chatushtaya*, or four modes of assimilation or process for reducing to a common denominator, fractions having dissimilar denominators, preliminary to addition and subtraction of fractions.

[3] *Bhága-játi* or *ansa-savarnana*, assimilation of fractions, reducing them to uniformity.

[4] *Bhága, ansa, vibhága, lava, &c.*, the numerator of a fraction. *Hara, hára, chheda, &c.*, the denominator of a fraction. That which is to be divided is the part (*ansa*); and that by which it is to be divided is *hara*, the divisor.—Gan. and Súr.

[5] *Rási*, a quantity, § 36.

[This is the ordinary rule for reducing fractions to their least common denominator. The first part of the rule is meant for fractions whose denominators are prime to each other.]

30. Example. Tell me the fractions reduced to a common denominator which answer to three and a fifth, and one-third, proposed for addition; and those which correspond to a sixty-third and a fourteenth offered for subtraction.

Statement[1] : $\frac{3}{1}, \frac{1}{5}, \frac{1}{3}$.

Reduced to a common denominator, $\frac{45}{15}, \frac{3}{15}, \frac{5}{15}$. Sum $\frac{53}{15}$.

Statement of the second example: $\frac{1}{63}, \frac{1}{14}$.

The denominators being abridged, or reduced to least terms, by the common measure 7, the fractions become $\frac{1}{9}, \frac{1}{2}$. Numerator and denominator, multiplied by the abridged denominators, give respectively $\frac{2}{126}$ and $\frac{9}{126}$. Subtraction being made, the difference is $\frac{7}{126}$. This abridged by 7 is $\frac{1}{18}$.

31. Rule for the reduction of subdivided fractions[2] : half a stanza.

---

[1] Among astronomers and other arithmeticians, oral instruction has taught to place the numerator above and the denominator beneath.—Gan.

No line is interposed in the original; but it has been introduced in the translation to conform to the modern practice. Bháskara subsequently directs (§ 36) an integer to be written as a fraction by placing under it unity for its denominator. The same is done by him in this place in the text.

[2] *Prabhága-játi*, assimilation of sub-fractions, or making uniform the fraction of a fraction.—Gan.

The numerators being multiplied by the numerators, and the denominators by the denominators, the result is a reduction to homogeneous form in subdivision of fractions.

[This is the ordinary rule for reducing a compound fraction to a simple fraction.]

32. Example. The quarter of a sixteenth of the fifth of three-quarters of two-thirds of a moiety of a *dramma* was given to a beggar by a person, from whom he asked alms: tell me how many cowry shells the miser gave, if thou be conversant, in arithmetic, with the reduction termed subdivision of fractions.

Statement: $\frac{1}{1}\ \frac{1}{2}\ \frac{2}{3}\ \frac{3}{4}\ \frac{1}{5}\ \frac{1}{16}\ \frac{1}{4}$.

Reduced to homogeneousness, $\frac{6}{7680}$, or in least terms, $\frac{1}{1280}$.

Answer: A single cowry shell was given.[1]

33. Rule for the reduction of quantities increased or decreased by a fraction:[2] a stanza and a half.

Prabhága, a divided fraction or fraction of a fraction: as a part of a moiety, and so forth.—Gang.

[1] For a cowry shell is in the tale of money the 1280th part of a *dramma*, § 2.

[2] *Bhágánubandhajáti*, assimilation of fractional increase, reduction to uniformity of an increase by a fraction, or the addition of a part; from *anubandha*, junction.—Gan. *Bhágápaváhajáti*, assimilation of fractional decrease; from *apaváha*, deduction.—Gan.

These, as remarked by Ganesa, are merely particular cases of addition and subtraction. The fractions may be parts of an integer, or parts of the proposed quantity itself. Hence we get two sorts of each, named

The integer being multiplied by the denominator, the numerator is made positive or negative,[1] provided parts of a unit be added or be subtractive. But if indeed the quantity be increased or diminished by a part of itself, then, in the addition and subtraction of fractions, multiply the denominator by the denominator standing underneath[2], and the numerator by the same augmented or lessened by its own numerator.

[The first part of the rule follows from the identity, $a \pm \frac{b}{c} = \frac{ac \pm b}{c}$, and the second part, from the identities,

$$\frac{a}{b} \pm \frac{c}{d} \times \frac{a}{b} = \frac{ad \pm ac}{bd} = \frac{a(d \pm c)}{bd}$$

Hence, if we write $\frac{c}{d}$ underneath $\frac{a}{b}$, we get the second part of the rule.

The process may be repeated, if necessary.

Thus, $\frac{a}{b} + \frac{c}{d} \times \frac{a}{b} + \frac{e}{f}\left(\frac{a}{b} + \frac{c}{d} \times \frac{a}{b}\right) = \frac{a(d+c)(f+e)}{bdf}$.

An application of this last formula occurs in the examples given in § 35.]

---

by Gangádhara and Súryadása, *rúpa-bhágánubandha*, addition of the fraction of a unit; *rúpa-bhágápaváha*, subtraction of the fraction of a unit; *rási-bhágánubandha*, addition of a fraction of the quantity; *rási-bhágápaváha*, subtraction of a fraction of the quantity.

[1] *Dhana*, positive; *rina*, negative.

[2] Indian arithmeticians write fractions under the quantities to which they are additive, or from which they are subtractive. Accordingly, the numerators and denominators are put in their order, one under the other.

34. Example. Say how much two and a quarter, and three less a quarter, are, when reduced to uniformity, if thou be acquainted with fractional increase or decrease.

Statement : $\begin{matrix} 2 & 3 \\ \frac{1}{4} & \cdot\frac{1}{4} \end{matrix}$.

Reduced to homogeneousness, they become $\frac{9}{4}$ and $\frac{11}{4}$.

[In the original a dot (·) is used instead of the sign *minus* (−).]

35. Example. How much is a quarter added to its third part, with a half of the sum, and how much are two-thirds lessened by one-eighth of them, and then diminished by three-sevenths of the residue? Tell me likewise, how much half less its eighth part, added to nine-sevenths of the residue is, if thou be skilled, dear woman, in fractional increase and decrease.

Statement : $\begin{matrix} \frac{1}{4}, & \frac{2}{3}, & \frac{1}{2} \\ \frac{1}{3}, & \cdot\frac{1}{8}, & \cdot\frac{1}{8} \\ \frac{1}{2}, & \cdot\frac{3}{7}, & \frac{9}{7}. \end{matrix}$

Reduced to uniformity, the results are $\frac{1}{2}$, $\frac{1}{3}$, $\frac{1}{1}$.

[In the above examples we may apply the last formula given in § 33. Thus in the first example, we have $a=1$, $b=4$, $c=1$, $d=3$, $e=1$, $f=2$, and all the signs are *plus*. Hence the result is $\frac{1 \times (3+1) \times (2+1)}{4 \times 3 \times 2} = \frac{12}{24} = \frac{1}{2}$; and similarly for the other two. The same process is exhibited in the *Manoranjana*.]

THE EIGHT RULES OF ARITHMETIC APPLIED TO FRACTIONS.[1]

36. Rule for addition and subtraction of fractions[2] : half a stanza. The sum or (in the case of subtraction) the difference of fractions having a common denominator, is (taken). Unity[3] is put denominator of a quantity[4] which has no divisor.[5]

[This rule and the others which follow are all ordinary rules.]

37. Example. Tell me, dear woman, quickly, how much a fifth, a quarter, a third, a half, and a sixth, make when added together. Say instantly what the residue of three is, subtracting those fractions.

Statement : $\frac{1}{5}$, $\frac{1}{4}$, $\frac{1}{3}$, $\frac{1}{2}$, $\frac{1}{6}$.

Added together the sum is $\frac{29}{20}$.

Subtracting those fractions from three, the remainder is $\frac{31}{20}$.

---

[1] *Bhinna-parikarmáshtaka,* the eight modes of process, as applicable to fractions : the preceding Section relating to those arithmetical processes as applicable to integers (*abhinna-parikarmáshtaka.*)

[2] *Bhinna-sankalita,* addition of fractions ; *bhinna-vyavakalita,* subtraction of fractions.

[3] *Rúpa,* the species or form ; anything having bounds.—Gang. In the singular, the arithmetical unit ; in the plural, any integer.

[4] *Rási,* a congeries ; a heap of things, of which unity is the scale of numeration ; a quantity or number.

[5] That is, it is put denominator of an integer.

38. Rule for multiplication of fractions[1] : half a stanza.

The product of the numerators, divided by the product of the denominators, (gives a quotient, which) is the result of multiplication of fractions.

39. Example. What is the product of two and a seventh, multiplied by two and a third, and of a moiety multiplied by a third? Tell, if thou be skilled in the method of multiplication of fractions.

Statement : 2 2
$\frac{1}{3}$ $\frac{1}{7}$.

Reducing to uniformity we get $\frac{7}{3}$, $\frac{15}{7}$. The product is $\frac{5}{1}$.

Statement : $\frac{1}{2}$ $\frac{1}{3}$.

The product is $\frac{1}{6}$.

40. Rule for division of fractions[2] : half a stanza.

After reversing the numerator and denominator of the divisor, the remaining process for division of fractions is that of multiplication.

41. Example. Tell me the result of dividing five by two and a third ; and a sixth by a third ; if thy understanding, sharpened into confidence, be competent for the division of fractions.[3]

---

[1] *Bhinna-gunana*, multiplication of fractions.

[2] *Bhinna-bhágahára*, division of fractions.

[3] Ganesa omits the latter half of the stanza. Gangádhara gives it entire.

Statement : $2\frac{1}{3}$ $(\frac{7}{3})$ $\frac{5}{1}$. $\frac{1}{3}$, $\frac{1}{6}$.

Proceeding as directed, the quotients are $\frac{15}{7}$ and $\frac{1}{2}$.

42. Rule for involution and evolution of fractions[1] : half a stanza.

If the square be sought, find both squares ; if the cube be required, both cubes : or, to discover the root (of cube or square), extract the roots of both (numerator and denominator.)

43. Example. Tell me quickly the square of three and a half ; and the square root of the square ; and the cube of the same ; and the cube root of that cube ; if thou be conversant with fractional squares and roots.

Statement : $3\frac{1}{2}$ or reduced $\frac{7}{2}$.

Its square is $\frac{49}{4}$; of which the square root is $\frac{7}{2}$. The cube of it is $\frac{343}{8}$; of which again the cube root is $\frac{7}{2}$.

SECTION IV.

CIPHER.[2]

44—45. Rule for arithmetical process relative to cipher : two couplets.

---

1 *Bhinna-varga*, square of a fraction ; *bhinna-ghana*, cube of a fraction.

2 *Śûnya*, *kha*, and other synonyms of vacuum or ethorial space ; nought or cipher ; a blank or the privation of specific quantity.—Krishna on *Vijaganita*.

In addition, cipher makes the sum equal to the additive.[1] In involution and (evolution) the result is cipher. A definite quantity,[2] divided by cipher, is the submultiple of nought.[3] The product of cipher is nought: but it must be retained as a multiple of cipher,[4] if any further operation impend. Cipher having become a multiplier, should nought afterwards become a divisor, the definite quantity must be understood to be unchanged. So likewise any quantity, to which cipher is added, or from which it is subtracted, (is unaltered).

[The first four rules are clear. The rule, *viz.*, "cipher having become a multiplier, &c," is not accurate. For $\frac{a \times 0}{0} = \frac{0}{0} =$ indeterminate, and, *not* $= a$, as the rule says. The idea of *infinity* is not introduced here by the author. It is, however, introduced by him in the *Vija-ganita*, and also by Ganesa in his commentary on the above couplets.]

46. Example. Tell me how much cipher added to five is, and the square of cipher, and its square

---

[1] *Kshepa*, that which is cast or thrown in; additive.—Gang.

[2] Rasi. See § 36.

[3] *Kha-hara*, a fraction with cipher for its denominator. According to the remark of Ganesa, it is an infinite quantity: since it cannot be determined how great it is. It remains unaltered by the addition or subtraction of finite quantities: since, in the preliminary operation of reducing both fractional expressions to a common denominator, preparatory to taking their sum or difference, both numerator and denominator of the finite quantity vanish. Ranganátha affirms that it is infinite, because the smaller the divisor is, the greater is the quotient: now cipher, being in the utmost degree small, given a quotient infinitely great.

[4] *Khaguna*, a quantity which has cipher for its multiplier. Cipher is set down by the side of the multiplicand, to denote it.—Gan.

root, its cube, and cube root ; and five multiplied by cipher ; and how much ten is subtracting cipher ; and what number it is, which multiplied by cipher, and added to half itself, and multiplied by three, and divided by cipher, amounts to a given number sixty-three.

Statement : 0. Cipher added to 5 makes 5. Square of cipher, 0. Square root, 0. Cube of cipher, 0. Cube root, 0.

Statement : 5. This multiplied by cipher makes 0.

Statement : 10. This divided by cipher gives $\frac{10}{0}$.

Statement : an unknown quantity ; its multiplier, 0 ; additive, $\frac{1}{2}$, multiplicator, 3 ; divisor, 0 ; given number, 63 ; assumption, 1.

Then, either by inversion or position, as subsequently explained (§47 and §50), the number is found, 14. This mode of computation is of frequent use in astronomical calculation.

[The last example as translated by Colebrooke appears to be meaningless and absurd. If we put x for the required number, we get the equation,

$$\frac{3\,(x \times 0 + \frac{1}{2}x)}{0} = 63,$$

which is manifestly absurd. The correct translation, however, would lead to the equation,

$$\frac{0 \times (x + \frac{1}{2}x) \times 3}{0} = 63,$$

of which $x = 14$ is a solution.]

CHAPTER III.

Miscellaneous Rules.[1]

SECTION I.

INVERSION.

47—48. Rule of inversion[2]: two stanzas. To investigate a quantity, one being given,[3] make the divisor a multiplicator; and the multiplier a divisor; the square, a root; and the root, a square[4]; turn the negative into positive, and the positive into negative. If a quantity is to be increased or diminished by its own proportionate part, let the (lower[5]) denominator, being increased or diminished by its numerator, become the (corrected[6]) denominator, and the numerator remain unchanged; and then proceed with the other operations of inversion, as before directed.

[The reason for the rule is clear from the example given in § 49. If we want an arithmetical solution of such a problem, we must begin from the end, and invert every operation indicated in the problem. If a quantity is to be increased or diminished by its own proportionate part, *i. e.*, if we have an equation of the form $x\left(1 \pm \frac{a}{b}\right) = c$, then

[1] *Prakírna*, miscellaneous. The rules contained in the first five sections of this chapter have none answering to them in the Arithmetic of Brahmagupta and Srídhara.

[2] *Viloma-vidhi*, *Viloma-kriyá*, *Vyasta-vidhi*, inversion.

[3] *Drisya*, the quantity or number, which is visible; the given quantity

[4] And the cube, a cube root; and the cube root, a cube.—Gan.

[5] Gangádhara

[6] Gangádhara.

evidently $x=\frac{bc}{b\pm a}=c\mp\frac{ac}{b\pm a}$ This explains the latter part of the rule.]

49. Example. Pretty girl, with tremulous eyes, if thou know the correct method of inversion, tell me the number, which multiplied by three, and added to three-quarters of the product, and divided by seven, and reduced by subtraction of a third part of the quotient, and then multiplied into itself, and having fifty-two subtracted from the product and the square root of the remainder extracted, and eight added, and the sum divided by ten, yields two.[1]

Statement: Multiplier 3. Additive $\frac{3}{4}$. Divisor 7. Decrease $\frac{1}{3}$. Square—. Subtractive 52. Square root—. Additive 8. Divisor 10. Given number 2.

Proceeding as directed, the result is 28, the number sought.

SECTION II.

SUPPOSITION.

50. Rule of supposition[2]: one stanza. Any number assumed at pleasure is treated as specified

---

[1] All the operations are inverted. The known number 2, multiplied by the divisor 10 converted into a multiplicator, makes 20; from which the additive 8, being subtracted, leaves 12; the square whereof (extraction of the root being directed) is 144; and adding the subtractive 52, it becomes 196; the root of this (square being directed) is 14; added to its half, 7, it amounts to 21, which multiplied by 7, is 147. This again divided by 7 and multiplied by 3 makes 63, which, subtracted from 147, leaves 84; and this divided by 3, gives 28.—*Mano.*

[2] *Ishta-karman,* operation with an assumed number. It is the rule of false position, supposition, and trial and error.

in the particular question, being multiplied and divided, raised or diminished by fractions; then the given quantity, being multiplied by the assumed number and divided by that (which has been found), yields the number sought. This is called the process of supposition.

51. Example. What is that number, which multiplied by five, and having the third part of the product subtracted, and the remainder divided by ten, and one-third, a half and a quarter of the original quantity added, gives two less than seventy?

Statement: Multiplier 5. Subtractive $\frac{1}{3}$ of itself. Divisor 10. Additive $\frac{1}{3}$ $\frac{1}{2}$ $\frac{1}{4}$ of the quantity. Given 68.

Putting 3: this multiplied by 5 is 15; less its third part, is 10: divided by 10, yields 1. Added to the third, half and quarter of the assumed number 3, *viz.*, $\frac{3}{3}$ $\frac{3}{2}$ $\frac{3}{4}$, the sum is $\frac{17}{4}$. By this divide the given number 68 taken into the assumed one 3; the quotient is 48.

The answer is the same with any other assumed number, as 1, &c.

Thus, by whatever number the quantity is multiplied or divided in any example, or by whatever fraction of the quantity it is increased or diminished, by the same should the like operations be performed on a number arbitrarily assumed; and by that, which results, divide the giver number taken into the assumed one; the quotient is the quantity sought.

[The rule in § 50 is a clumsy way of solving a simple equation. The reason for it will appear below.

Let x denote the number sought in § 51.

Then, $5x \times \frac{2}{3} \times \frac{1}{10} + (\frac{1}{3} + \frac{1}{2} + \frac{1}{4})x = 68$.

Multiply both sides by any assumed integer k, and we get $x = \{68 \times k\} \div \{k \times 5 \times \frac{2}{3} \times \frac{1}{10} + (\frac{1}{3} + \frac{1}{2} + \frac{1}{4})k\}$.

Thus we see that there is no need at all of assuming an integer k, as the rule directs.]

52. Example of reduction of a given quantity.[1] Out of a heap of pure lotus flower, a third part, a fifth and a sixth were offered respectively to the gods Siva, Vishnu and the Sun; and a quarter was presented to Bhaváni. The remaining six lotuses were given to the venerable preceptor. Tell quickly the whole number of lotuses.

Statement: $\frac{1}{3}, \frac{1}{5}, \frac{1}{6}, \frac{1}{4}$; known 6. Putting one for the assumed number, and proceeding as above, the quantity is found 120.

[In Pandit Jívánanda Vidyáságara's edition, there is an example before this, which is omitted by Colebrooke. It is as follows :—

Out of a herd of elephants, half together with a third part of itself was roaming in a forest; a sixth part together with a seventh of itself was drinking water in a river; and an eighth part together with a ninth of itself was playing with lotuses. The leader of the herd was seen accompanied by three females. What was the number of elephants in the herd?

---

[1] *Drisyá-játi*, reduction of the visible or given quantity with fractions affirmative or negative; here, with negative; in the preceding example with affirmative.

This may be solved either by an arithmetical or an algebraical method, both being practically the same. Adopting the latter, and putting x for the required number, we get the equation $\{\frac{1}{2}(1+\frac{1}{3})+\frac{1}{6}(1+\frac{1}{7})+\frac{1}{8}(1+\frac{1}{9})\}x+4=x$, whence, $x = 1008$.

It may be observed here that all the examples in this Section are problems producing simple equations, which are solved not by the ordinary method of solving simple equations, but by the author's method stated in § 50. They may also be worked out by a purely arithmetical method.

The algebraical solution of the problem in § 52 is as follows :—

Let x denote the whole number of lotuses.

Then $(\frac{1}{3}+\frac{1}{5}+\frac{1}{6}+\frac{1}{4})x+6 = x$, whence $x = 120$].

53. Example of reduction of residues.[1] A traveller, engaged in a pilgrimage, gave half his money at *Prayága*; two-ninths of the remainder at *Kásí*; a quarter of the residue in payment of taxes on the road; six-tenths of what was left at *Gayá*; there remained sixty-three *nishkas*, with which he returned home. Tell me the amount of his original stock of money, if you have learned the method of reduction of fractions of residues.

Statement: $\frac{1}{2}, \frac{2}{9}, \frac{1}{4}, \frac{6}{10}$; known 63. Putting one for the assumed number, subtracting the numerator from its denominator, multiplying denominators together, and in other respects proceding as directed, the remainder is found $\frac{7}{60}$. By this

[1] *Sesha-játi*, assimilation of residue; reduction of fractions of residues or successive fractional remainders.

dividing the given number 63 taken into the assumed quantity, the original sum comes out 540.

Or it may be found by the method of reduction of fractional decrease (§ 33).

Statement : $\frac{1}{1}.\frac{1}{2}.\frac{2}{9}.\frac{1}{4}.\frac{6}{10}$. Being reduced to homogeneous form, the result is $\frac{7}{60}$: whence the sum is deduced 540.

Or this may also be found by the rule of inversion (§ 47).

[Let x denote the original stock in *nishkas*. Then, at *Prayága*, there remained $\frac{1}{2}x$; at *Kási*, $\frac{2}{9}$ of this being spent, there remained $\frac{7}{9}$ of $\frac{1}{2}x$; similarly, on the road, there remained $\frac{3}{4}$ of $\frac{7}{9}$ of $\frac{1}{2}x$, and at *Gayá*, $\frac{4}{10}$ of $\frac{3}{4}$ of $\frac{7}{9}$ of $\frac{1}{2}x$ or $\frac{7}{60}x$; hence we get $\frac{7}{60}x = 63$, whence $x = 540$.]

54. Example of reduction of differences.[1] Out of a swarm of bees, one-fifth part settled on a blossom of *Kadamba*,[2] and one-third on a flower of *silindhri*[3]; three times the difference of those numbers flew to the bloom of a *Kutaja*.[4] One bee, which remained, hovered and flew about in the air, allured at the same moment by the pleasing fragrance of a jasmine and pandanus. Tell me, charming woman, the number of bees.

Statement : $\frac{1}{5}, \frac{1}{3}, \frac{2}{15}$; known quantity, 1 ; assumed, 30.

A fifth of the assumed number is 6 ; a third is 10 ; difference 4 ; multiplied by 3 gives 12 ; and the

---

[1] *Vislesha-játi*, assimilation of difference ; reduction of fractional differences.

[2] *Kadamba*, Nauclea orientalis or N. Kadamba.

[3] *Silindhri*, a plant resembling the *Kachora*.—Krishna or *Vija-ganita*

[4] Echites antidysenterica.

remainder is 2. Then the product of the known quantity by the assumed one, being divided by this remainder shows the number of bees 15.

Here also putting unit for the assumed quantity, the number of the swarm is found 15.

So in other instances likewise.[1]

[Let x denote the number of bees.

Then, $\frac{1}{5}x+\frac{1}{3}x+3(\frac{1}{3}x-\frac{1}{5}x)+1 = x$, whence $x = 15$.]

SECTION III.

55. Rule of concurrence: half a stanza.

The sum with the difference added and subtracted, being halved, gives the two quantities. This is termed concurrence.[2]

[Let x and y denote the required numbers. Then, $x+y=k$, $x-y=l$, where k and l are given quantities; whence $x = \frac{1}{2}(k+l)$, $y = \frac{1}{2}(k-l)$.]

56. Example. Tell me the numbers, the sum of which is a hundred and one, and the difference,

---

[1] The *Manoranjana* introduces one more example, which is there placed after the second, and is here subjoined :—"The third part of a necklace of pearls, broken in an amorous struggle, fell to the ground; its fifth part rested on the couch; the sixth part was saved by the wench; and the tenth part was taken up by her lover; six pearls remained strung. Say of how many pearls the necklace was composed."

Statement : $\frac{1}{3}$ $\frac{1}{5}$ $\frac{1}{6}$ $\frac{1}{10}$. Rem. 6.

Answer : 30.

[2] *Sankrámana*, concurrence or mutual penetration in the shape of sum and difference.—Gang. Investigation of two quantities concurrent or grown together in the form of sum and difference.—Gan. Calculation of quantities latent within those exhibited.—Súr.

twenty five; if thou know the rule of concurrence, dear child.

Statement: sum 101; difference 25.

The two numbers are 38 and 63.

57. Rule of dissimilar operation[1]; half a stanza.

The difference of the squares, divided by the difference of the radical quantities, gives their sum[2]; whence the quantities are found in the mode before directed.

[Let x and y be the numbers. Then,

$$x^2 - y^2 = m, \; x - y = n.$$

$$\therefore \quad x + y = \frac{m}{n}, \text{ \&c.}]$$

58. Example. Tell me quickly, skilful calculator, what numbers they are, of which the difference is eight, and the difference of squares four hundred.

Statement: difference of the quantities 8; difference of the squares 400.

The numbers are 21 and 29.

SECTION IV.

POBLEM CONCERNING SQUARES.[3]

A certain problem relative to squares is propounded in the next instance.

---

[1] *Vishama-karman*, the finding of the quantities, when the difference of their squares is given, and either the sum or the difference of the quantities.—Gan. A species of concurrence.—Gang. See below, §135.

[2] Or divided by their sum, gives their difference.—Gan.

[3] *Varga-karman*, operation relative to squares; an indeterminate problem, admitting innumerable solutions.

59. Rule. The square of an arbitrary number, multiplied by eight and lessened by one, then halved and divided by the assumed number, is one quantity; its square, halved and added to one, is the other. Or unity, divided by double an assumed number and added to that number, is a first quantity; and unity is the other. These give pairs of quantities, the sum and difference of whose squares, lessened by one, are squares.

[Let n be the assumed number. Then, by the first part of the rule, the two numbers are,

$$\frac{1}{2n}(8n^2-1) \text{ and } \tfrac{1}{2}\left\{\frac{1}{2n}(8n^2-1)\right\}^2+1$$

The sum of the squares of these numbers lessened by 1 is

$$\tfrac{1}{4}\left(4n-\frac{1}{2n}\right)^4+2\left(4n-\frac{1}{2n}\right)^2$$

$$=\left(4n-\frac{1}{2n}\right)^2\left\{\tfrac{1}{4}\left(4n-\frac{1}{2n}\right)^2+2\right\}$$

$$=\left(4n-\frac{1}{2n}\right)^2\left(2n+\frac{1}{4n}\right)^2, \text{ a perfect square.}$$

Similarly, the difference of the squares of the numbers lessened by 1 is a perfect square.

Again, by the second part of the rule, the numbers are $\frac{1}{2n}+n$ and 1; and $\left\{\left(\frac{1}{2n}+n\right)^2 \pm (1)^2\right\}-1=\left(\frac{1}{2n}\pm n\right)^2$, which are perfect squares.

Hence the reason for the rule is evident.]

60. Tell me, my friend, numbers, the sum and difference of whose squares, less one, afford square roots, which dull smatterers in algebra labour to

excruciate, puzzling for it in the six-fold method of discovery there taught.[1]

To bring out an answer by the first rule, let the number put be $\frac{1}{2}$. Its square, $\frac{1}{4}$, multiplied by 8, is 2; which lessened by 1 is 1. This halved is $\frac{1}{2}$, and divided by the assumed number $\frac{1}{2}$ gives 1 for the first quantity. Its square halved is $\frac{1}{2}$, which, added to 1, makes $\frac{3}{2}$. Thus the two quantities are 1 and $\frac{3}{2}$.

So, putting 1 for the assumed number, the numbers obtained are $\frac{7}{2}$ and $\frac{57}{8}$. With the supposition of 2, they are $\frac{31}{4}$ and $\frac{993}{32}$.

By the second method, let the assumed number be 1. Unity divided by the double of it is $\frac{1}{2}$, which added to the assumed number makes $\frac{3}{2}$. The first quantity is thus found. The second is unity. With the supposition of 2, the quantities are $\frac{9}{4}$ and 1. Putting 3, they are $\frac{19}{6}$ and 1.

61. Another Rule.[2] The square of the square of an arbitrary number, and the cube of that number respectively multiplied by eight, adding one to the first product, are such quantities, equally in arithmetic and algebra.

---

1 This question, found in some copies of the text, and interpreted by Gangádhara and the *Manoranjana*, is unnoticed by the other commentators. [We do not know what the author means by *the six-fold method of discovery*. Colebrooke does not say anything about it.—Ed.]

2 To bring out answers in whole numbers, the two preceding solutions giving fractions.—Gan. and Súr.

Put $\frac{1}{2}$. The square of the square of the assumed number is $\frac{1}{16}$, which multiplied by 8 makes $\frac{1}{2}$. This added to 1 is $\frac{3}{2}$, which is the first quantity. Again put $\frac{1}{2}$. Its cube is $\frac{1}{8}$, which multiplied by 8 gives the second quantity 1. Next supposing 1, the two quantities are 9 and 8. Assuming 2, they are 129 and 64. Putting 3, they are 649 and 216. And so on, without end, by means of various suppositions, in the several foregoing methods.

It is said that algebraic solution similar to arithmetical rules appears obscure; but it is not so to the intelligent; nor is it six-fold, but manifold.

[Let n be the arbitrary number. Then, by the rule, the numbers are, $8n^4+1$ and $8n^3$;
and $\{(8n^4+1)^2 \pm (8n^3)^2\} - 1 = (4n^2)^2(2n^2 \pm 1)^2$, which are perfect squares.

Hence the reason for the rule is obvious.]

SECTION V.

62—63. Rule for assimilation of the root's coefficient[1]: two stanzas.

The sum or difference of a quantity and of a multiple of its square root being given, the square of half the coefficient[2] is added to the given number, and the square root of their sum (is extracted; that root,) with half the coefficient added or subtracted,

---

[1] *Múla-játi*, *múlá-gunaka-játi* or *ishta-múlánsa-játi*, assimilation and reduction of the root's coefficient with a fraction.

[2] *Guna*, multiplicator; *múla-guna*, root's multiplier, the coefficient of the root.

being squared, is the quantity sought by the interrogator. If the quantity have a fraction (of itself) added or subtracted, divide the number given and the multiplicator of the root, by unity increased or lessened by the fraction, and the required quantity may be then discovered, proceeding with those quotients as above directed.

A quantity, increased or diminished by its square root multiplied by some number, being given, add the square of half the multiplier of the root to the given number; and extract the square root of the sum. Add half the multiplier, if the difference were given; or subtract it, if the sum were so. The square of the result will be the quantity sought.

[The third paragraph is in prose in the original, and is added by the author by way of explanation of the two preceding metrical rules.

Suppose we have the equations,

$$x \pm a\sqrt{x} = b \qquad \ldots \qquad \ldots \qquad (1)$$

Then, completing the square, we get

$$x \pm a\sqrt{x} + \left(\frac{a}{2}\right)^2 = b + \left(\frac{a}{2}\right)^2;$$

$$\therefore \sqrt{x} \quad \sqrt{b + \left(\frac{a}{2}\right)^2} \mp \frac{a}{2};$$

$$\therefore x = \left\{\sqrt{b + \left(\frac{a}{2}\right)^2} \mp \frac{a}{2}\right\}^2.$$

Hence the reason for the first part of the rule is clear. It is the ordinary rule for solving an equation reducible to a quadratic by completing the square.

The second part of the rule is meant for equations of the form

$$x \pm \frac{c}{d}x \pm a\sqrt{x} = b \quad \ldots \quad \ldots \quad (2).$$

Whence we get

$$x \pm \frac{a\sqrt{x}}{1 \pm \frac{c}{d}} = \frac{b}{1 \pm \frac{c}{d}},$$

which is of the form (1), and may be solved as above. Thus we see the reason for the second part of the rule.]

64. Example: the root subtracted, and the difference given. One pair out of a flock of geese remained sporting in the water, and saw seven times the half of the square root of the flock proceeding to the shore tired of the diversion. Tell me, dear girl, the number of the flock.

Statement: coeff. $\frac{7}{2}$; given 2. Half the coefficient is $\frac{7}{4}$; its square $\frac{49}{16}$ added to the given number, makes $\frac{81}{16}$, the square root of which is $\frac{9}{4}$. Half the coefficient being added, the sum is $\frac{16}{4}$; or, reduced to least terms, 4. This squared is 16; the number of the flock, as required.

[Let x denote the number of the flock.

Then, $2+\frac{7}{2}\sqrt{x} = x$;

$$\therefore\ x-\frac{7}{2}\sqrt{x}+\left(\frac{7}{4}\right)^2 = 2+\left(\frac{7}{4}\right)^2 = \left(\frac{9}{4}\right)^2;$$

$$\therefore\ \sqrt{x}-\frac{7}{4} = \frac{9}{4};$$

whence $x = \left(\frac{7}{4}+\frac{9}{4}\right)^2 = 16.$

This is an instance of the first part of the preceding rule.]

65. Example: the root added, and the sum given. Tell me what the number is, which, added

to nine times its square root, amounts to twelve hundred and forty.

Statement: coeff. 9; given 1240. Proceeding by the rule, the required number is 961.

[Let x denote the number required. Then, $x+9\sqrt{x}=1240$, whence x].

66. Example: the root and a fraction both subtracted. Of a flock of geese, ten times the square root of the number departed for the *Mánasa* lake,[1] on the approach of a cloud: and eighth part went to a forest of *Sthalapadminís*[2]: three couples were seen engaged in sport on the water abounding with delicate fibres of the lotus. Tell, dear girl, the whole number of the flock.

Statement: coeff. 10; fraction $\frac{1}{8}$; given 6.

Proceeding by the (second) rule §(63), unity less the fraction, is $\frac{7}{8}$; and the coefficient and the given number, being both divided by that, become $\frac{80}{7}$ and $\frac{48}{7}$; and the half coefficient is $\frac{40}{7}$. With these, proceeding by the (first) rule §(62), the number of the flock is found 144.

[Let x denote the whole number of the flock.

Then, $10\sqrt{x}+\frac{1}{8}x+6=x$;

---

[1] Wild geese are observed to quit the plains of India, at the approach of the rainy season; and the lake called *Mánasasarovara* is covered with water-fowl, especially geese, during that season. The *Hindus* suppose the whole tribe of geese to retire to the holy lake at the approach of rain. The bird is sacred to Brahmá. [See *Raghuvansa*, XIII, 55.—Ed.]

[2] The plant intended is not ascertained. The context would seem to imply that it is arboreous, as the term signifies forest.

$\therefore\ \frac{7}{8}x-10\sqrt{x}=6$; whence x.

This is an instance of the second part of the rule in §62—63.]

67. Example. The son of Prithá,[1] irritated in fight, shot a quiver of arrows to slay Karna. With half his arrows, he parried those of his antagonist; with four times the square root of the quiver-full, he killed his horses; with six arrows, he slew Salya[2]; with three he demolished the umbrella, standard and bow; and with one, he cut off the head of the foe. How many were the arrows which Arjuna let fly?

Statement: fraction $\frac{1}{2}$; coeff. 4; given 10.

The given number and coefficient being divided by unity less the fraction become 20 and 8; and proceeding by the rule (§62), the number of arrows comes out 100.

[Let x denote the number of arrows.

Then, $\frac{1}{2}x+4\sqrt{x}+6+3+1=x$, whence x.]

68. Example. The square root of half the number of a swarm of bees is gone to a shrub of jasmin[3]; and so are eight-ninths of the whole swarm: a female is buzzing to one remaining male that is humming within a lotus in which he is confined, having been allured to it by its fragrance at night.[4] Say, lovely woman, the number of bees.

---

[1] Arjuna, surnamed Pártha; his matronymic from Prithá or Kunti.

[2] One of the *Kauravas*, and charioteer of Karna.

[3] *Málati*, jasminum grandiflorum.

[4] The lotus being open at night and closed in the day, the bee might be caught in it.—Gan.

Here eight-ninths of the quantity and the root of its half are negative (and cosequently subtractive) from the quantity: and the given number is two of the specific things. The negative quantity, and the given number halved, bring out half the quantity sought.[1] Thus :—

Statement: fraction $\cdot\frac{8}{9}$; coeff. $\cdot\frac{1}{2}$; given 1.

A fraction of half the quantity is the same as half the fraction of the quantity; the fraction is therefore set down (unaltered).

Here proceeding as above directed, there comes out half the quantity, 36; which being doubled is the number of bees in the swarm, 72.

[Let x denote the number of bees.

Then, $\sqrt{\frac{1}{2}x}+\frac{8}{9}x+2=x$.

Put $y=\frac{1}{2}x$, and we get

$$y-\frac{8}{9}y-\frac{1}{2}\sqrt{y}=1,$$

whence by the rule in §§62-63, we obtain $y=36$, and $\therefore$ $x=72$.

Thus the reason for the process given in the text is clear. The reason given by the author and the commentators is not very clear.]

69. Example : the root and a fraction both added. Find quickly, if thou have skill in arith-

[1] In such questions it is necessary to observe whether the coefficient of the root be so of the root of the whole number, or of that of its part; for that quantity is found, of whose root the coefficient is used. But in the present case, the root of half the quantity is proposed; and accordingly, the half of the quantity will be found by the rule. The number given, however, belongs to the entire quantity. Therefore, taking half the given number, half the required number is to be brought out by the process before directed,—*Mano*, and Súr.

metic, the quantity which added to its third part and eighteen times its square root, amounts to twelve hundred.

Statement : fraction $\frac{1}{3}$; coeff. 18 ; given 1200.

Here, dividing the coefficient and given number by unity added to the fraction §63), and proceeding as before directed, the number is brought out, 576.

[Let x denote the number required.

Then, $x+\frac{1}{3}x+18\sqrt{x}=1200$; whence x.]

[The following problems leading to quadratic equations are given by Bháskara in the *Vijaganita*:—

The eighth part of a troop of monkeys, squared, was skipping in a grove and delighted with their sport. Twelve remaining were seen on the hill, amused with chattering to each other. How many were they in all ? †

Putting x for the required number of monkeys, we get the equation $\left(\frac{x}{8}\right)^2+12=x$, whence $x=16$ or 48, both values being admissible.

The fifth part of a troop of monkeys less three, being squared, had gone to a cave ; and one monkey was in sight, having climbed on a branch. Say how many they were. ‡

---

† वनान्तराले प्लवगाष्टभागः संवर्गितो वल्गतिजातरागः ।

फूत्कारनादप्रतिनादहृष्टा दृष्टा गिरौ द्वादश ते कियन्तः ॥

‡ यूथात्पञ्चांशकस्त्र्यूनो वर्गितो गह्वरंगतः ।

दृष्टः शाखामृगः शाखामारूढो वदते कति ॥

Putting x for the number of monkeys, we get the equation $\left(\frac{x}{5}-3\right)^2+1=x$, whence $x=50$ or 5, the latter value being inadmissible. This is noticed by Bháskara who says, "But the second (value) is in this case not to be taken, for it is incongruous. People do not approve a negative absolute number".]

SECTION VI.

Rule of Proportion.[1]

70. Rule of three terms[2] : one stanza.

The first and last terms, which are the argument and requisition, must be of like denomination ; the fruit, which is of a different species, stands between them : and that, being multiplied by the demand and divided by the first term, gives the fruit of the demand.[3] In the inverse method, the operation is reversed.[4]

[The rule is the ordinary mechanical one for solving problems involving the Rule of Three, direct and inverse. It is not stated in the light of the principle of proportion, and is practically the same as that given in Mr. Barnard Smith's well-known work on Arithmetic, Art. 155.]

---

1 [A more literal translation would be 'Rule of Three,' the word in the original being *trairásika*.—Ed.]

2 *Trairásik*, calculation belonging to a set of three terms.—Gang. Rule of Three. The first term is *pramána*, measure or argument; the second is its fruit, *phala*, or produce of the argument ; the third is *ichchhá*, the demand, requisition, desire or question.—Gan.

3 *Ichchhá-phala*, produce of the requisition, or fruit of the question ; it is of the same denomination or species with the second term.

4 See § 74.

71. Example. If two and a half *palas* of saffron be obtained for three-sevenths of a *nishka*, say instantly, best of merchants, how much is got for nine *nishkas*.

Statement : $\frac{3}{7}$ $\frac{5}{2}$ $\frac{9}{1}$. Answer: 52 *palas* and 2 *karshas*.

[This is an example of the Rule of Three direct. Worked out by the principle of proportion, the process will stand thus:—

Let x denote the quantity sought in *palas*.

Then, we evidently have the proportion,

$$\tfrac{5}{2} : x :: \tfrac{3}{7} : 9,$$

$$\therefore \quad x \times \tfrac{3}{7} = \tfrac{5}{2} \times 9,$$

$$\therefore \quad x = \frac{\frac{5}{2} \times 9}{\frac{3}{7}} = 52\tfrac{1}{2}.$$

Thus the answer is $52\frac{1}{2}$ *palas* = 52 *palas*, 2 *karshas*.

The reason for the rule in § 70 is obvious].

72. Example. If one hundred and four *nishkas* are got for sixty-three *palas* of best camphor, consider and tell me, friend, what may be obtained for twelve and a quarter *palas*.

Statement: 63 104 $\frac{49}{4}$. Answer: 20 *nishkas*, 3 *drammas*, 8 *panas*, 3 *kákinís*, 11 cowryshells and $\frac{1}{9}$th part.

[This also is an instance of the Rule of Three direct, and may be worked out as above.]

73. Example. If a *khári* and one-eighth of rice may be procured for two *drammas*, say quickly what may be had for seventy *panas*.

Statement, reducing *drammas* to *panas*: 32 $\frac{9}{8}$ 70.

Answer: 2 *khàris*, 7 *dronas*, 1 *ádhaka*, 2 *prasthas*.

[This is a third instance of the Rule of Three direct.]

74. Rule of Three inverse[1].

If the fruit diminish as the requisition increases, or augment as that decreases, they who are skilled in accounts consider the Rule of Three terms to be inverted.[2]

When there is diminution of fruit, if there be increase of requisition, and increase of fruit if there be diminution of requisition, then the inverse Rule of Three is (employed).

[This is the ordinary definition of inverse variation.]

75. For instance, when the value of living beings[3] is regulated by their age; and in the case of gold, where the weight and touch are compared[4], or when heaps[5] are subdivided; let the inverted Rule of Three terms be (used).

---

[1] *Vyasta-trairásika* or *Viloma-trairásika*. rule of three terms inverse,

[2] The method of performing the inverse rule has been already taught (§70), *viz.*, "in the inverse method, the operation is reversed;" *i.e.*, the fruit is to be multiplied by the argument and divided by the demand.—Súr.

When the fruit increases or decreases, as the demand is augmented or diminished, the direct rule (*Krama-trairásika*) is used; else the inverse.—Gan.

[3] Slaves and cattle. The price of the older is less; of the younger, greater.—Gang. and Súr.

[4] Colour on the touchstone. See Alligation, § 101.

[5] *See* Chap. X. When heaps of grain, which had been meted with a small measure, are again meted with a larger one, the number decreases; and when those, which had been meted with a large measure, are again meted with a smaller one, there is increase of number.—Gang. and Súr.

[Some instances of inverse variation are here mentioned. The reason is clear from the foot-notes appended. The author does not mention the common instance of time and agency required for a given piece of work.]

76. Example of age and price of living beings. If a female slave, sixteen years of age, bring thirty-two (*nishkas*), what will one aged twenty cost? If an ox which has been worked two years sell for four *nishkas*, what will one, which has been worked six years, cost?

1st Statement: 16 32 20. Answer: $25\frac{3}{5}$ *nishkas*

2nd Statement: 2 4 6. Answer: $1\frac{1}{3}$ *nishka*.

[Let x denote the cost in the first example. Then $\because$ the greater is the age, the smaller is the cost, we have the proportion,

$$16 : 20 :: x : 32,$$

$$\text{Whence } x = \frac{16 \times 32}{20} = 25\tfrac{3}{5}.$$

Similarly the second example as well as those in the next two articles may be worked out.]

77. Example of touch and weight of gold. If a *gadyánaka* of gold of the touch of ten may be had for one *nishka* (of silver), what weight of gold of fifteen touch may be bought for the same price?

Statement: 10 1 15. Answer $\frac{2}{3}$.

78. Example of subdivision of heaps. A heap of grain having been meted with a measure containing seven *ádhakas*, if a hundred such measures were found, what would be the result with one containing five *ádhakas*?

Statement: 7 100 5. Answer 140.

79. Rule of compound proportion[1]: one stanza.

In the method of five, seven, nine or more[2] terms, transpose the fruit and divisors[3]; and the product of the larger set of terms, being divided by the product of the less set of terms[4], the quotient is the produce sought).

[This is practically the same rule, rather incompletely stated, for solving problems involving the Double Rule of

---

[1] [A more literal translation would be, "Rule of Five and so forth," the word in the original being, *pancharásikádau*.—Ed.]. This, which is the compound Rule of Three, comprises, according to Ganesa, two or more sets of three terms (*trairásika*); or two or more proportions (*anupáta*), as Súryadása observes. Thus the rule of five (*pancha-rásika*) comprises two proportions; that of seven (*sapta-rásika*) three; that of nine (*nava-rásika*) four; and that of eleven (*ekádasa-rásika*) five.

[2] Meaning eleven. *Mano*, and Súr.

[3] Ganesa and the commentator of the *Vásaná* understand this last word (*chhid*, divisor) as relating to denominators of fractions; and the transposing of them (if any there be) is indeed right: accordingly the author gives under this rule an example of working with fractions (§ 81). But the *Manoranjana* and Súryadása explain it otherwise; and the latter cites as, ancient commentary entitled *Ganita-Kaumudi* in support of his exposition. "There are two sets of terms; those which belong to the argument, and those which appertain to the requisition. The fruit in the first set is called produce of the argument; that in the second is named divisor of the set. They are to be transposed, or reciprocally brought from one set to the other; *i.e.*, put the fruit in the second set, and the divisor in the first. Would it not be enough to say, transpose the fruits of both sets? The author of the *Kaumudi* replies, 'the designation of divisor serves to indicate that after transposition, the fruit of the second set being included in the product of the less set of terms, the product of the greater set is to be divided by it'. Some, however, interpret it as relative to fractions. But that is wrong: for the word would be superfluous." This explanation is not very clear.—Ed.]

[4] *Bahu-rási-paksha*, set of many terms. That to which the fruit is brought is the larger set.—Gang. Or, if there be fruit on both sides, that, in which the fruit of the requisition is, is the larger set.—Gan. *Laghu-rási-paksha*, set of fewer terms.

Three, as that given in Mr. Barnard Smith's work on Arithmetic, Art. 161.* It is a clumsy and a purely mechanical rule, having no connection whatever with the principle of proportion. The meaning of the phrase, "transpose the fruit and divisors," as explained by Ganesa, will appear from the foot-notes appended to the following articles. It should be observed here that "the product of the larger set" is not necessarily the numerically larger product; see the example in §82. For the meaning of the term "larger set," see foot-note 3 p. 48. The phrase has been rather loosely used.]

80. Example. If the interest of a hundred for a month be five, say what the interest of sixteen is for a year. Find likewise the time from the principal and interest; and knowing the time and produce, tell the principal sum.

Statement:	1	12
	100	16
	5	

Answer[1]: the interest is $9\frac{3}{5}$.

To find the time; statement:	1	
	100	16
	5	$\frac{48}{5}$

Answer[2]: months 12.

---

[1] Transposing the fruit,

1	12
100	16
	5

Product of the larger set, 960. Do. of the less set, 100. Quotient, $\frac{960}{100}$ or $\frac{48}{5}$.

[2] Transposing both fruits,

1	
100	16
$\frac{48}{5}$	5

and the denominator,

1	
100	16
48	5
	5

Product of the larger set, 4800. Do. of the less set, 400. Quotient, 12.

To find the principal; statement: $\begin{matrix} 1 & 12 \\ 100 & \\ 5 & \frac{48}{5} \end{matrix}$

Answer[1] : principal 16.

[Worked out by the principle of proportion, the process will stand thus :—

Let x denote the interest required. Then, the int. of 16 for 1 year = int. of 16×12 for 1 month; and $\because$ with a given time, the interest varies directly as the principal, we get the proportion, $100 : 16 \times 12 : : 5 : x$,
whence $x = \frac{16 \times 12 \times 5}{100} = 9\frac{3}{5}$,
and the reason for the rule in § 79 is evident. Similarly for the other parts of the example.]

81. Example. If the interest of a hundred for a month and one-third be five and one-fifth, say what the interest is of sixty-two and a half for three months and one-fifth.

Statement: $\begin{matrix} \frac{4}{3} & \frac{16}{5} \\ 100 & \frac{125}{2} \\ \frac{26}{5} & \end{matrix}$ Answer[1] : interest $7\frac{4}{5}$.

[This may be worked out similarly as the preceding example.]

---

[1] Transposing both fruits, $\begin{matrix} 1 & 12 \\ 100 & \\ \frac{48}{5} & 5 \end{matrix}$ and the denominator, $\begin{matrix} 1 & 12 \\ 100 & \\ 48 & 5 \\ & 5 \end{matrix}$

Product of the larger set, 4800.
Do. of the less set 300. Quotient, 16.

[2] Transposing the fruit, $\begin{matrix} \frac{4}{3} & \frac{16}{5} \\ 100 & \frac{125}{2} \\ & \frac{26}{5} \end{matrix}$ and the denominators, $\begin{matrix} 4 & 16 \\ 5 & 3 \\ 100 & 125 \\ 2 & \\ 5 & 26 \end{matrix}$

Abridging by correspondent reduction on both sides, $\begin{matrix} 1 & 4 \\ 5 & 3 \\ 4 & 5 \\ 2 & \\ 5 & 26 \end{matrix}$ and by

82. Example of the Rule of Seven. If eight best variegated silk scarfs, measuring three cubits in breadth and eight in length, cost of a hundred (*nishkas*); say quickly, merchant, if thou understand trade, what a like scarf, three and a half cubits long and half a cubit wide, will cost.

Statement:

3	$\frac{1}{2}$
8	$\frac{7}{2}$
8	$\frac{1}{1}$
100	

Answer[1] : *Nishka* 0, *drammas* 14, *panas* 9, *kákiní* 1, cowry-shells $6\frac{2}{3}$.

[Let x denote the cost required.]

Then the area of cloth in the first case

$=8\times3\times8$ sq. cubits;

and in the second case

$=1\times\frac{7}{2}\times\frac{1}{2}$ sq. cubits.

Hence, the quality of the cloth remaining the same, we get the proportion,

$$8\times3\times8 : 1\times\frac{7}{2}\times\frac{1}{2} :: 100 : x,$$

whence $x=\frac{1\times7\times1\times100}{8\times3\times8\times2\times2}$,

---

further reduction,

1	1
1	3
1	1
1	
5	13

Product of the larger set, 39. Do. of the less set, 5. Quotient, $7\frac{4}{5}$.

The abridgment of the work by reduction of terms on both sides by their common divisors is taught by the *Manoranjana*.

[1] Transposing fruit and denominators

3	1
2	
8	7
2	
8	1
	100

Product of the larger set, 700. Do. of the less set 768. Quotient : 0, 14, 9, 1, $6\frac{2}{3}$.

and the reason for the process given in the foot-note is evident].

83. Example of the Rule of Nine. If thirty benches, twelve fingers thick, square of four wide, and fourteen cubits long, cost a hundred (*nishkas*); tell me, my friend, what price fourteen benches will fetch, which are four less in every dimension.

Statement:	12	8	Answer[1] : *nishkas* $16\frac{2}{3}$.
	16	12	
	14	10	
	30	14	
	100		

[Here, putting x for the price, and proceeding as above, we get the proportion,

$30 \times 12 \times 16 \times 14 \times 24 : 14 \times 8 \times 12 \times 10 \times 24 :: 100 : x$,

whence x is known.]

84. Example of the Rule of Eleven. If the hire of carts to convey the benches of the dimensions first specified (in the preceding example), through a distance of one league (*gavyúti*[2]) be eight *drammas*; say what the cart-hire should be for bringing the benches last mentioned, four less in every dimension, through a distance of six leagues.

---

[1] Transposing fruit.

12	8	Abridging by	1	1
16	12		2	1
14	10	correspondent	1	1
30	14	reduction on	3	1
	100	both sides.		100

Product of the larger set, 100. Quotient, $16\frac{2}{3}$.
Do. of the less set, 6.

[2] *Gavyúti*, two *krosas* or half a *yojana*; it contains rather more than 8000 yards, and is more than $4\frac{1}{2}$ English miles.

Statement:

12	8
16	12
14	10
30	14
1	6
8	

Answer[1] : *drammas* 8

[Let x denote the cart-hire required.

Then in the first case, solid content of benches $= 30 \times 12 \times 16 \times 14 \times 24$ cubic fingers, and they are carried through a distance of 1 league; and in the second case, solid content of benches $= 14 \times 8 \times 12 \times 10 \times 24$ cubic fingers, and they are carried through a distance of 6 leagues, which is the same as a solid content of $14 \times 8 \times 12 \times 10 \times 24 \times 6$ cubic fingers carried through 1 league;
hence, since for a given distance, the hire will vary directly as the solid content of benches carried, we get the proportion,
$30 \times 12 \times 16 \times 14 \times 24 \times 1 : 14 \times 8 \times 12 \times 10 \times 24 \times 6 :: 8 : x$,
whence $x = \frac{14 \times 8 \times 12 \times 10 \times 6 \times 8}{30 \times 12 \times 16 \times 14} = 8$.]

85. Rule of barter[2]; half a stanza. So in barter

---

[1] Transposing the fruit,

12	8
16	12
14	10
30	14
1	6
	8

Abridging by correspondent reduction on both sides,

1	1
2	1
1	1
3	1
1	6
	8

and by further reduction,

1	1
1	1
1	1
1	1
1	2
	4

Product of the larger set, 8.
Do. of the less set, 1.
Quotient, 8.

[2] *Bhānda-prati-bhāndaka*, commodity for commodity; computation of the exchange of goods (*vastu-vinimaya-ganita*,—Gang.): barter.

likewise, the same process is (followed); transposing both prices, as well as the divisors.[1]

[The reason for the rule will appear from the example which follows].

86. Example. If three hundred mangoes be had in the market for one *dramma*, and thirty ripe pomegranates for a *pana*; say quickly, friend, how many should be had in exchange for ten mangoes.

Statement:

16	1
300	30
10	

Answer[2]: 16 pomegranates.

[The example involves two proportions, as Súryadása observes.

---

[1] Gangádhara, Súryadása and the *Manoranjana* so read this passage, *haránscha-múlye*. But Ganesa and Ranganátha have the affirmative adverb *sadáhi* (always) in place of the word *haránscha* (and the divisors). At all events the transposition of denominators takes place, as usual; and so does that of the lowest term or fruit, as in the Rule of Five, to which, as Súryadása, remarks, this is analogous. It comprises two proportions, thus stated by him from the example in the text:—"If for one *pana*, thirty pomegranates may be had, how many for sixteen? Answer, 480. Again, if for 300 mangoes, 480 pomegranates may be had, how many for ten? Answer, 16. Here thirty is first multiplied by sixteen and then divided by one; and then multiplied by ten and divided by three hundred. For brevity, the prices are transposed, and the result is the same."

[2] Transposing the prices,

1	16
300	30
10	

and transposing the fruit,

1	16
300	30
	10

Product of the larger set, 4800.
Do. of the less set, 300. Quotient, 16.

First find how many pomegranates can be had for one *dramma* or 16 *panas*.

Putting y for this number, we get

$$30 : y :: 1 : 16,$$

$$\therefore\ y = 30 \times 16.$$

Hence by the question, 300 mangoes are equivalent to 30×16 pomegranates.

Then, putting x for the number of pomegranates required, we get the proportion,

$$300 : 10 :: 30 \times 16 : x.$$

$$\text{whence } x = \frac{30 \times 16 \times 10}{300} = 16,$$

and the reason for the rule in § 85 is obvious.]

CHAPTER IV.

INVESTIGATION OF MIXTURE.[1]

SECTION I.

INTEREST.

87—88. Rule : a stanza and a half.[2]

The argument[3] multiplied by its time, and the fruit multiplied by the mixed quantity's time, being severally set down, and divided by their sum and multiplied by the mixed quantity, are the principal and interest (composing the quantity). Or the principal being found by the rule of supposition (§ 50), that, taken from the mixed quantity, leaves the amount of interest.

[The rule refers specially to the example given in § 89. By the word *argument* is meant 100, and by the word *fruit* is meant the interest on 100 for 1 month, or, as we would call it, the rate per cent. per mensem.

Let r = rate per cent. per mensem.
t = time in years.
P = principal.
I = interest.
A = amount.

---

[1] *Misra-vyavahára*, investigation of mixture, ascertainment of composition, as principal and interest, and so forth.—Gaṇ. It is chiefly grounded on the rule of proportion.—*Ibid.* The rules in this chapter bear reference to the examples which follow them. Generally they are *quæstiones otiosæ*; problems for exercise.

[2] To investigate the principal and interest, the amount, time and rate being given.—Gan.

[3] *Pramána*, argument; *phala*, fruit (§ 70) : principal and interest.

Then, $I = \frac{P \times r \times 12 \times t}{100}$.

$$\therefore \quad A = P\left\{1 + \frac{r \times 12 \times t}{100}\right\} = P \times \frac{100 + r \times 12 \times t}{100}.$$

$$\therefore \quad P = \frac{A \times 100}{100 + r \times 12 \times t}, \text{ and } I = \frac{A \times r \times 12 \times t}{100 + r \times 12 \times t}.$$

The last two formulæ stated in words give the first part of the rule. It is evident that these formulæ may also be derived by a simple proportion, as observed by Ganesa.

In the latter part of the rule, the principal is found by the author's method of supposition, which is practically the same as the solution of a simple equation.]

89. Example. If the principal sum, with interest at the rate of five on the hundred by the month, amount in a year to one thousand, tell the principal and interest respectively.

Statement: 1 12
100 1000.
5

Answer[1] : principal, 625 ; interest, 375.

Or, by the rule of position, put one ; and proceeding according to that rule (§50), the interest of unity is $\frac{3}{5}$, which, added to one, makes $\frac{8}{5}$. The given quantity 1000, multiplied by unity and divided by that ($\frac{8}{5}$), shows the principal 625. This,

[1] 100 multiplied by 1 is 100 ; 5 by 12 is 60. Their sum 160 is the divisor. The first number 100, multiplied by 1000, and divided by 160, is 625. The second 60, multiplied by 1000, and divided by 160, gives 375.—Gang.

8

taken from the mixed amount, leaves the interest[1] 375.

[The second solution follows the latter part of the preceding rule.

Let x denote the principal required. Then ∵ the rate per cent. per annum is 60, we get

$$x\left(1+\frac{60}{100}\right) = 1000,$$

$$\text{whence} \quad x = \frac{1000}{1+\frac{3}{5}}.\ ]$$

90. Rule[2] : The arguments taken into their respective times are divided by the fruit taken into the elapsed times ; the several quotients, divided by their sum and multiplied by the mixed quantity, are the parts as severally lent.

[The rule refers specially to the example in § 91.

Let x, y, z be the portions lent at r_1, r_2, r_3 per cent. per mensem, and let I = common interest in t_1, t_2, t_3 months respectively.

Then, $x+y+z = a$, a given quantity ;

$$\text{and } \frac{x\times r_1\times t_1}{100} = \frac{y\times r_2\times t_2}{100} = \frac{z\times r_3\times t_3}{100} = I$$

$$\therefore \quad x : y : z :: \frac{100\times 1}{r_1\times t_1} : \frac{100\times 1}{r_2\times t_2} : \frac{100\times 1}{r_3\times t_3} ;$$

$$\therefore \quad x = \frac{100\times 1}{r_1\times t_1}\times\frac{a}{\frac{100\times 1}{r_1\times t_1}+\frac{100\times 1}{r_2\times t_2}+\frac{100\times 1}{r_3\times t_3}},$$

---

[1]Or the principal being known, the interest may be found by the Rule of Five.—Sūr.

[2]For determining parts of a compound sum.—Sūr.

with similar values for y and z. Hence the rule. Ganesa's explanation is practically the same as the above, but it is rather obscure : see foot note 1.]

91. Example. The sum of six less than a hundred *nishkas* being lent in three portions at interest of five, three and four per cent., an equal interest was obtained on all three portions, in seven, ten and five months respectively. Tell, mathematician, the amount of each portion.[1]

Statement :

1	7	1	10	1	5
100		100		100	
5		3		4	

Mixed amount 94.

Answer[2] : the portions are 24, 28 and 42. The equal amount of interest is $8\frac{2}{5}$.

---

[1] Since the amount of interest on all the portions is the same, put unity for its arbitrarily assumed amount : whence corresponding principal sums are found by the Rule of Five. For instance, if a hundred be the capital, of which five is interst for a month, what is the capital, of which unity is the interest for seven months ? and in like manner, the principal sums are to be found. Thus, a compound proportion being wrought, the time is multiplied by the argument to which it appertains, and divided by the fruit taken into the elapsed time. Then, as the total of those principal sums is to them severally, so is the given total to the respective portions lent. They are thus severally found by the Rule of Three.—Gan.

[2] Multiplying the argument and fruit by the times, and dividing one product by the other, there result the fractions $\frac{100}{35}$, $\frac{100}{30}$, $\frac{100}{20}$; or $\frac{20}{7}$, $\frac{20}{6}$, $\frac{20}{4}$; which reduced to a common denominator and summed, make $\frac{1880}{168}$. Multiplied by the mixed amount 94, the fractions become $\frac{1880}{7}$, $\frac{1880}{6}$, $\frac{1880}{4}$; and then divided by the sum $\frac{1880}{168}$, they give $\frac{168}{7}$, $\frac{168}{6}$. $\frac{168}{4}$, or 24 28, 42.—*Mano.*

To find the interest, employ the Rule of Five ; $\begin{matrix} 1 & \\ 100 & 24 \\ 5 & 7 \end{matrix}$ Answer. $8\frac{2}{5}$

By the same method, with all three portions, the interest comes out the same.—Súr.

92. Rule[1] : half a stanza.

The contributions[2] being multiplied by the mixed amount and divided by the sum of the contributions, are the respective fruits.[3]

[If we divide the mixed amount into parts proportional to the contributions, we shall get the mixed amount as due to each contribution. And this amount less the contribution is the gain. Hence the reason for the rule is obvious.]

93. Example. Say, mathematician, what the apportioned shares are of three traders, whose original capitals were respectively fifty-one, sixty-eight and eighty-five, which have been raised by commerce conducted by them on joint stock, to the aggregate amount of three hundred.

Statement : 51, 68, 85 ; sum, 204. Mixed amount, 300.

Answer : 75, 100, 125. These, less the capital sums, are 1 the gains : *viz.*, 24, 32, 40.

Or the mixed amount, less the sum of the capitals, is the profit on the whole : *viz.*, 96. This being

---

[1] The capital sums, their aggregate amount, and the sum of the gains being given, to apportion the gains.—Gan.

[2] *Prakshepaka*, that which is thrown in or mixed.—Gan. Joined together.—Súr.

[3] The principle of the rule is obvious, being simply the Rule of Three. —Gan.

If by this sum of contributions, this contribution be had, then by the compound sum what will be ? The numbers thus found, less the contributions. are the gains.—Ranganátha on the *Vásaná*.

multiplied by the contributions and divided by their sum, gives the respective gains: *viz.*, 24, 32, 40.

[The whole gain being divided into parts proportional to the contributions, gives the respective gains.]

SECTION II.

94. Rule[1]: half a stanza. Divide denominators by numerators; and then divide unity by those quotients added together. The result will be the time of filling (a cistern by several fountains).

[The rule refers to an example of the class given in § 95.

Let the times in which the fountains can severally fill the cistern be $\frac{a_1}{b_1}$, $\frac{a_2}{b_2}$, &c., of a day. Then in one day, the parts of the cistern filled up by the fountains severally are $\frac{b_1}{a_1}$, $\frac{b_2}{a_2}$, &c.; $\therefore$ if they work together, the part filled in one day will be $= \frac{b_1}{a_1} + \frac{b_2}{a_2} + \&c.$;

$\therefore$ the time in which the whole cistern will be filled

$$= \frac{1}{\frac{b_1}{a_1} + \frac{b_2}{a_2} + \&c.} \text{ of a day.}$$

whence the reason for the rule is evident. The explanations of this rule given by the commentators are all vague and unsatisfactory.]

95. Example. Say quickly, friend, in what portion of a day (four) fountains, being let loose

---

[1] To apportion the time for a mixture of springs to fill a well or cistern.—Gan. To solve an instance relative to fractions.—Súr.

together, will fill a cistern, which, if severally opened, they would fill in one day, half a day, the third, and the sixth part, respectively.

Statement: $\frac{1}{1}$, $\frac{1}{2}$, $\frac{1}{3}$, $\frac{1}{6}$.

Answer: $\frac{1}{12}$th part of a day.

SECTION III.

PURCHASE AND SALE.

96. Rule.[1] By the (measure of the) commodities,[2] divide their prices taken into their respective portions (of the purchase); and by the sum of the quotients divide both them and those portions severally multiplied by the mixed sum: the prices and quantities are found in their order.

[The reason for the rule will appear from the solution of the example in § 97.]

97. Example. If three and a half *mánas*[3] of rice may be had for one *dramma*, and eight of kidney-beans[4] for the like price, take these thirteen *kákinís*, merchant, and give me quickly two

[1] For a case where a mixture of portions and composition of things are given.—Gan. Concerning measure of grain, &c.—Súr.

[2] *Panya*, the measure of the grain or other commodity procurable for the current price in the market.—Súr, and the *Mano*

[3] *Mána* or *mánaka*, a measure: seemingly intending a particular one. According to Ganesa, the *mána* (apparently the same as the *máni*) is at most an eighth of a *khári*: being a cubic span. See note to § 236. A spurious couplet (see note on § 2) makes it the modern measure of weight containing forty *sers*.

Mudga, phaseolus mungo: a sort of kidney-bean.

parts of rice with one of kidney-beans ; for we must make a hasty meal and depart, since my companion will proceed onwards.

Statement : $\frac{2}{1}$ $\frac{1}{1}$. Mixed sum $\frac{13}{64}$.
$\frac{1}{1}$ $\frac{1}{1}$
$\frac{7}{2}$ $\frac{8}{1}$

The prices, $\frac{1}{1}$, $\frac{1}{1}$, multiplied by the portions, $\frac{2}{1}$, $\frac{1}{1}$, and divided by the goods $\frac{7}{2}$, $\frac{8}{1}$, make $\frac{4}{7}$, $\frac{1}{8}$, the sum of which is $\frac{39}{56}$. By this divide the same fractions $\frac{4}{7}$, $\frac{1}{8}$, taken into the mixed sum $\frac{13}{64}$; and the portions $\frac{2}{1}$, $\frac{1}{1}$, taken into that mixed sum $\frac{13}{64}$. There result the prices of the rice and kidney-beans, $\frac{1}{6}$ and $\frac{7}{192}$ of a *dramma* ; or 10 *kákinís* and $13\frac{1}{3}$ shells for the rice, and 2 *kákinís* and $6\frac{2}{3}$ shells for the kidney-beans ; and the quantities are $\frac{7}{12}$ and $\frac{7}{24}$ of a *mána* of rice and kidney-beans respectively. .

[Let x denote the *mánas* of kidney-beans.
Then $2x$ will denote the *mánas* of rice.
Now the price paid $= \frac{13}{64}$ of a *dramma* ;

$$\therefore \quad 2x \times \tfrac{2}{7} + x \times \tfrac{1}{8} = \tfrac{13}{64} ;$$

$$\text{Whence} \quad x\left(\tfrac{4}{7}+\tfrac{1}{8}\right) = \tfrac{13}{64},$$

$$\therefore \quad x = \frac{1 \times \frac{13}{64}}{\frac{4}{7}+\frac{1}{8}} = \tfrac{7}{24},$$

$$\text{and} \quad 2c = \frac{2 \times \frac{13}{64}}{\frac{4}{7}+\frac{1}{8}} = \tfrac{7}{12} ;$$

$$\text{and the prices are,} \quad \tfrac{7}{24} \times \tfrac{1}{8} = \tfrac{7}{192}$$

$$\text{and} \quad \tfrac{7}{24} \times \tfrac{4}{7} = \tfrac{1}{6}.$$

Hence the reason for the rule is evident. General formulæ corresponding to the rule in § 96 may be easily established. It is, however, not worth while to do so.

The example in § 98 may be worked out in a similar manner.]

98. Example. If a *pala* of best comphor may be had for two *nishkas*. and a *pala* of sandal-wood[1] for the eighth part of a *dramma*, and half a *pala* of alæ-wood[2] also for the eighth of a *dramma*, good merchant, give me the value of one *nishka* in the proportions of one, sixteen and eight ; for I wish to prepare a perfume.

Statement : 32 $\frac{1}{8}$ $\frac{1}{8}$. Mixed sum 16.

1 1 $\frac{1}{2}$

1 16 8

Answer. Prices : *drammas*, $14\frac{2}{9}$, $\frac{8}{9}$, $\frac{8}{9}$.

Quantities : *palas*, $\frac{4}{9}$, $7\frac{1}{9}$, $3\frac{5}{9}$.

SECTION IV.

99. Rule. Problem concerning a present of gems.[3]

From the gems subtract the gift multiplied by the persons ; and any arbitrary number being divided by the remainders, the quotients are numbers expressive of the prices. Or the remainders being multiplied together, the product, divided

---

[1] *Chandana* ; santalum album.

[2] *Aguru* ; aquillaria agallochum.

[3] The problem is an indeterminate one. The solution gives relative values only.

by the several reserved remainders, gives the values in whole numbers.

[The reason for the rule will appear from the solution of the example in § 100, to which the rule specially refers.]

100. Example. Four jewellers, possessing respectively eight rubies, ten sapphires, a hundred pearls, and five diamonds, presented, each from his own stock, one apiece to the rest in token of regard and gratification at meeting ; and they thus became owners of stock of precisely equal value. Tell me severally, friend, the prices of their gems.

Statement : rub. 8 ; sapph. 10 ; pearls 100 ; diam. 5 ; gift 1 ; persons 4.

Here, the product of the gift 1 by the persons 4, *viz.*, 4, being severally subtracted, there remain rubies 4, sapphires 6, pearls 96, diamond 1. Any number arbitrarily assumed being divided by these remainders, the quotients are the relative values. Taking it at random, they may be fractional values ; or by judicious selection, whole numbers. Thus, put 96 ; and the prices thence deduced are 24, 16, 1, 96 ; and the equal stock 233.

Or the remainders being multiplied together, and the product severally divided by those remainders, the prices are 576, 384, 24, 2304 ; and the equal amount of stock (after interchange of presents) is 5592.

[Let the relative values of a ruby, sapphire, pearl, and diamond be respectively x, y, z, w. Then we shall evidently get from the conditions of the problem the following equations :—

$$\begin{aligned} &5x+y+z+w \\ &=7y+x+z+w \\ &=97z+x+y+w \\ &=2w+x+y+z \end{aligned}$$

$\therefore 4x=6y=96z=w=k$ suppose; then $x=\frac{k}{4}, y=\frac{k}{6}, z=\frac{k}{96}, w=k$.

Putting k=L. C. M. of 4, 6, 96, *i. e.*, 96, we get the least integral values of x y, z, w, *viz.*, 24, 16, 1, 96; and putting k=product of 4, 6, 96, we get for x, y, z, w, the values 576, 384, 24, 2304.

The reason for the rule in § 99 will be evident from the above algebraical solution. The coefficients of x, y, &c. in the final equations will be

=the no. of the respective gems—the no. of that gem given to each person × the no. of persons altogether.

For instance, if there were 9 rubies, and each presented 2 from his pocket to each of the rest, the coefficient of x in the left-hand side of the first equation would have been 3; and that of x in the right-hand side would have been 2; thus the coefficient of x in the simplified equations would have been 1, *i.e.*, $9-4\times 2$, which agrees with the rule in § 99.

Súryadása cites the *Víja-ganita* for the solution of the problem. Ranganátha gives an arithmetical explanation which, however, is not quite clear.

SECTION V.

ALLIGATION.[1]

101. Rule.[2] The sum of the products of the touch[3] and (weight of several parcels)[4] of gold being divided by the aggregate of the gold, the touch of the mass is found : or (after refining) being divided by the fine gold, the touch is ascertained ; or divided by the touch, the quantity of purified gold is determined.

[This is simply the ordinary rule for alligation medial. We may consider the prices per *másha* of the several kinds of gold as *proportional to* the fineness. The reason for the rule is obvious.]

102—103. Example. Parcels of gold weighing severally ten, four, two and four *máshas*, and of the fineness of thirteen, twelve, eleven and ten respectively, being melted together, tell me quickly, merchant, who art conversant with the computation of gold, what the fineness of the mass is. If the twenty *máshas* above described be reduced to sixteen by refining, tell me instantly the touch of

---

[1] *Suvarna-ganita,* computation of gold, that is, of its weight and fineness ; alligation medial.

[2] To find the fineness produced by mixture of parcels of gold ; and, after refining, to find the weight, if the fineness be known ; and the fineness if tho weight of refined gold be given.—Gan.

[3] *Varna,* colour of gold on the touchstone ; fineness of gold determined by that touch. See § 77. "The degrees of fineness increase as the weight is roduced by refining."—Gan.

Gang.

the purified mass. Or, if its purity when refined be sixteen, prithee, what is the number to which the twenty *máshas* are reduced ?

Statement: touch 13 12 11 10 ;

weight 10 4 2 4.

Answer[1] : after melting; fineness 12 ; weight 20.

After refining, the weight being sixteen, *máshas*, the touch is 15. The touch being sixteen, the weight is 15.

104. Rule.[2] From the acqured fineness of the mixture, taken into the aggregate quantity of gold, subtract the sum of the products of the weight and fineness (of the parcels, the touch of which is known), and divide the remainder by the quantity of gold of unknown fineness ; the quotient is the degree of its touch.[3]

105. Example. Eight *máshas* of ten, and two of eleven by the touch, and six of unknown fineness, being mixed together, the mass of gold, my friend, became of the fineness of twelve; tell the degree of unknown fineness.

Statement : 10 11 Fineness of the mixture 12

8 2 6.

---

1 Products 130, 48, 22, 40. Their sum 240, divided by 20, gives 12 ; divided by 16, gives 15.

2 To discover the fineness of a parcel of unknown degree of purity mixed with others of which the touch is known.—Gan.

3 The rule being the converse of the preceding, the principle of it is obvious.—Rang.

Answer : degree of the unknown fineness 15.

[Let x denote the unknown fineness. Then, $(8+2+6)\times 12=8\times 10+2\times 11+6\times x$,

whence $x=\dfrac{(8+2+6)\times 12-(8\times 10+2\times 11)}{6}$. The reason for the rule in § 104 is obvious.]

106. Rule.[1] The acquired fineness of the mixture being multiplied by the sum of the gold (in the known parcels), subtract therefrom the aggregate products of the weight and fineness (of the parcels) : divide the remainder by the difference between the fineness of the gold of unknown weight and that of the mixture, the quotient is the weight of gold that was unknown.

107. Example. Three *máshas* of gold of the touch of ten, and one of the fineness of fourteen, being mixed with some gold of the fineness of sixteen, the degree of purity of the mixture, my friend, is twelve. How many *máshas* are there of the fineness of sixteen ?

Statement: 10 14 16. Fineness of the mixture 12.
3 1

Answer : *másha* 1.

[Let x denote the number of *máshas* required.

Then, $(3+1+x)\times 12=3\times 10+1\times 14+x\times 16$;

$$\therefore\ x=\frac{(3+1)\times 12-(3\times 10+1\times 14)}{16-12},$$

whence the rule in § 106.]

---

[1] To find the weight of a parcel of known fineness, but unknown weight, mixed with other parcels of known weight and fineness.—Gan.

108. Rule.[1] Subtract the effected fineness from that of the gold of a higher degree of touch, and that of the one of lower touch from the effected fineness ; the differences, multiplied by an arbitrarily assumed number, will be the weights of gold of the lower and higher degrees of purity respectively.

109. Example. Two ingots of gold, of the touch of sixteen and ten respectively, being mixed together, the gold became of the fineness of twelve. Tell me, friend, the weight of gold in both lumps.

Statement : 16, 10. Fineness resulting 12.

Putting one, and proceeding as directed, the weights of gold are found, *máshas* 2 and 4. Assuming two, they are 4 and 8. Taking half, they come out 1 and 2. Thus, manifold answers are obtained by varying the assumption.

[Let x and y be the weights required.

Then, $x \times 16 + y \times 10 = (x+y) \times 12$;

$$\therefore \ (16-12)\,x = (12-10) \times y \ ;$$

$$\therefore \ \frac{x}{y} = \frac{12-10}{16-12} \ ;$$

$$\therefore \ x = (12-10)\,k, \ y = (16-12)k,$$

where k is any positive quantity.

The general solution in positive integers evidently is, $x = k$, $y = 2k$, where k is any positive integer.

The reason for the rule in § 108 is obvious.]

---

[1] To find the weight of two parcels of given fineness and unknown weight,—Gan. and Súr. The problem is an indeterminate one, as is intimated by the author.

SECTION VI.

PERMUTATIONS AND COMBINATIONS.

110—112. Rule[1] : three stanzas.

Let the figures from one upwards, differing by one, put in the inverse order, be divided by the same (arithmeticals) in direct order ; and let the subsequent be multiplied by the preceding, and the next following by the foregoing (result). The several results are the changes, ones, twos, threes, &c.[2] This is termed a general rule.[3] It serves in prosody, for those versed therein, to find the variations of metre ; in the arts (as in architecture) to compute the changes upon apertures(of a building); and (in music) the scheme of musical permutations[4] ; in medicine, the combinations of different savours. For fear of prolixity, this is not (fully) set forth.

[The reason for the rule will appear from the solution of the example which follows.]

[1] To find the possible permutations of long and short syllables in prosody ; combinations of ingredients in pharmacy ; variations of notes, &c., in music ; as well as changes in other instances.—Gan.

[2] According to Ganesa, there is no demonstration of the rule, besides acceptation and experience. [This, however, is not correct.—Ed.]

[3] Commentators appear to interpret this as a name of the rule here taught ; *sádhárana*, or *sádhárana-chhandoganita*, general rule of prosodian permutation, subject to modification in particular instances, as in music, where a special method (*asádhárana*) must be applied.—Gang. and Súr.

[4] *Khanda-meru*, a certain scheme.—Gan. It is more fully explained by other commentators ; but the translator is not sufficiently conversant with the theory of music to understand the term distinctly.

113. A single example in prosody. In the permutations of the *Gáyatrì* metre,[1] say quickly, friend, how many the possible changes of the verse are; and tell severally, how many the permutations are with one, (two, three,) &c., long syllables.

Here the verse of the *Gáyatrí* stanza comprises six syllables. Wherefore, the figures from one to six are set down, and the statement of them, in direct and inverse order is $\begin{matrix} 6 & 5 & 4 & 3 & 2 & 1 \\ 1 & 2 & 3 & 4 & 5 & 6 \end{matrix}$. Proceeding as directed, the results are :—changes with one long syllable, 6 ; with two, 15 ; with three, 20 ; with four, 15 ; with five, 6 ; with six, 1 ; with all short, 1. The sum of these is the whole number of permutations of the verse, 64.

In like manner, setting down the numbers of the whole tetrastic, in the mode directed, and finding changes with one, two, &c., and summing them, the permutations of the entire stanza are found, *viz.*, 16777216.

In the same way may be found the permutations of all varieties of metre, from *Ukthá* (which

---

[1] The *Gáyatrí* metre in sacred prosody is a triplet comprising twenty-four syllables, as in the famous prayer containing the Brahmnical creed, called *Gáyatrí* [*Rigveda, Mandala* 3, *súkta* 62, *rik.* 10.—Ed.] See As. Res., vol. X, p. 463. But in the prosody of profane poetry, the same number of syllables is distributed in a tetrastic ; and the verse consequently contains six syllables. (As. Res., vol. X, p. 469.)

consists of monosyllabic verses) to *Utkriti* (the verses of which contain twenty-six syllables).[1]

[In the *Gáyatri* metre, the number of syllables is 6. In finding the number of changes with *one* long and the rest *five* short syllables, we have to find the permutations of 6 things taken all at a time, when one of them is of one kind, and the rest, of another kind. Hence the number of changes $= \frac{\lfloor\underline{6}}{\lfloor\underline{1}\,\lfloor\underline{5}}$, which is precisely the number of combinations of 6 things taken 1 at a time. Similarly, in finding the number of changes with *two* long, and therefore the rest *four* short, we get the number $= \frac{\lfloor\underline{6}}{\lfloor\underline{2}\,\lfloor\underline{4}}$, and so on; thus finally the total number of changes = sum of combinations of 6 things taken 1, 2, 3, 4, 5, 6, at a time + 1 (with all short syllables) $=(2 - 1)+1=64$.

Now $\frac{\lfloor\underline{6}}{\lfloor\underline{1}\,\lfloor\underline{5}} = \frac{6}{1}$; $\frac{\lfloor\underline{6}}{\lfloor\underline{2}\,\lfloor\underline{4}} = \frac{6 \times 5}{1 \times 2}$; &c.

Hence the reason for the rule is clear.

If the aggregate number of changes only is wanted, this can be found at once from the proposition, *viz.*, the total number of combinations of n things taken 1, 2, 3..... n at a time $=2^n-1$ (see Todhunter's Algebra, art. 515). This proposition is given in a concrete shape in § 130–131.

Similarly, taking the whole tetrastic, *i.e.*, 24 syllables, the total number of changes $=(2^{24}-1)+1$ (with all short syllables) $=16777216$.]

[1] As. Res., vol. X, pp. 468–473.

114. Example. In a pleasant, spacious and elegant edifice, with eight doors,[1] constructed by a skilful architect, as a palace for the lord of the land, tell me the permutations of apertures taken one, two, three, &c.[2] Say, mathematician, how many are the combinations in one composition, with ingredients of six different tastes, sweet, pungent, astringent, sour, salt and bitter,[3] taking them by ones, twos, threes, &c.

Statement, first example :

8 7 6 5 4 3 2 1

1 2 3 4 5 6 7 8

Answer : the number of ways in which the doors may be opened by ones, twos, threes, &c., is 8, 28, 56, 70, 56, 28, 8, 1, respectively. And the changes on the apertures of the octagonal palace[4] amount to 255.

Statement, second example :

6 5 4 3 2 1

1 2 3 4 5 6

Answer : the number of various preparations with ingredients of divers tastes is 6, 15, 20, 15, 6, 1.[5]

---

[1] *Múshá*, aperture for the admission of air ; a door or window ; (same with *gaváksha*—Gan.) A portico or terrace, (*bhúmi visesha.*—Gang. and Súr.)

[2] The variations of one window or portico open (or terrace unroofed) and the rest closed ; two open, and the rest shut ; and so forth.

[3] *Amara-kosha, swarga-varga*, 147.

[4] An octagonal building, with eight doors or windows or porticos or terraces facing the eight cardinal points of the horizon, is meant—Gan.

[5] Total number of possible combinations is 63.—Gang.

[In the first example, the total number of variations $=2^8-1=255$. The case of all the windows being shut is not taken into account; otherwise the total number of variations would be 256.

The second example from its very nature is a case of combinations and not of permutations, *i.e.*, we have to find the number of combinations of 6 things taken 1, 2,...6 at a time. The rule in § 110—112, however, equally applies, as has been explained above. The total number of combinations in this case $= 2^6 - 1 = 63$.]

CHAPTER V.

PROGRESSIONS.[1]

SECTION I.

ARITHMETICAL PROGRESSION.

115. Rule.[2] Half the period[3] multiplied by the period added to unity, is the sum of the arithmeticals one, &c., and is named their addition.[4] This, being multiplied by the period added to two, and being divided by three, is the aggregate of the additions.[5]

---

[1] *Sredhi*, a term employed by the older authors for any set of distinct substances or other things put together.—Gan. It signifies sequence or progression. *Sredhi-vyavahára*, ascertainment or determination of progressions.

[2] To find the sum of the arithmeticals.—Gan.

[3] *Pada*, the place.—Gan. Any one of the figures or digits, being that of which the sum is required.—Súr. The last of the numbers to be summed.—*Mano.* See below, note to 119.

[4] *Sankalita*, the first sum or addition of arithmeticals. *Sankalitaikya*, aggregate of additions, summed sums or second sum.

[5] The first figure is unity. The sum of that and the period being halved, is the middle figure. As the figures decrease behind it, so they increase before it : wherefore the middle figure, multiplied by the period, is the sum of the figures one, &c., continued to the period. The only proof of the rule for the aggregate of sums is acceptation.—Gan. [This last remark is not correct. The proof of the formula for the sum of the first n arithmeticals given by Ganesa does not apply where n is even, but requires to be modified. In that case, the sum of any two terms equidistant from the first and last $= n+1$; whence the sum of n terms evidently $= \frac{n}{2}(n+1)$.—Ed.] It is a maxim, that a number multiplied by the next following arithmetical, and halved, gives the sum of the preceding;

[1+2+3......to n terms $=\frac{n(n+1)}{2}$. By "the aggregate of the additions," the author evidently means the sum of n terms of the series whose nth term is $\frac{1}{2}n(n+1)$, in other words, the sum of n *triangular numbers.* This sum is $\frac{1}{6}n(n+1)(n+2)=\frac{n(n+1)}{2}\cdot\frac{n+2}{3}$. See Todhunter's Algebra, Art. 666. The reason for the rule is obvious.]

116. Example. Tell me quickly, mathematician, the sums of the several (progressions of) numbers one, &c., continued to nine; and the summed sums of those numbers.

Statement: arithmeticals: 1 2 3 4 5 6 7 8 9.

Answer: sums: 1 3 6 10 15 21 28 36 45.

Summed sums: 1 4 10 20 35 56 84 120 165.

117. Rule.[1] Twice the period added to one and divided by three, being multiplied by the sum (of the arithmeticals), is the sum of the squares. The sum of the cubes of the numbers one, &c., is pronounced by the ancients equal to the square of the addition.

$$[1^2+2^2+\ldots\ldots+n^2=\frac{n(n+1)(2n+1)}{6}$$

$$=\frac{n(n+1)}{2}\cdot\frac{2n+1}{3}.$$

$$1^3+2^3+\ldots\ldots+n^3=\left\{\frac{n(n+1)}{2}\right\}^2.$$ See Todhunter's Algebra, Arts. 460, 461.]

---

wherefore, &c.—Súr. Kamalákara is quoted by Ranganátha for a demonstration grounded on placing the numbers of the series in the reversed order under the direct one and adding the two series—[the same proof as that given in modern works on Algebra.—Ed.]

[1] To find the sums of squares and of cubes.—Gan. and Súr.

118. Example. Tell promptly the sum of the squares, and the sum of the cubes, of those numbers, if thy mind be conversant with the way of summation.

Statement : 1 2 3 4 5 6 7 8 9.

Answer : sum of squares, 285. Sum of cubes, 2025.

119. Rule.[1] The increase multiplied by the period less one, and added to the first quantity, is the amount of the last.[2] That, added to the first, and halved, is the amount of the mean ; which multiplied by the period is the amount of the whole, and is denominated (*ganita*) the computed sum.

[Consider the series $a,\ a+b,\ a+2b$......The nth term $=a+(n-1)b$. If n be *odd*, middle term $=\frac{n+1}{2}$th term $=\frac{2a+(n-1)b}{2}$. If n be *even*, there will be two middle terms, *viz.*, $\frac{n}{2}$th and $\frac{n+2}{2}$th terms, and the mean amount = average of these two terms $=\frac{n}{2}$th term + $\frac{1}{2}$ of common difference

---

[1] Where the increase is arbitrary.—Gang. In such cases, to find the last term, mean amount, and sum of the progression.—Súr. From first term, common difference and period, to find the whole amount, &c.—Gan.

[2] *Adi* and *mukha*, *vadana*, *vaktra*, and other synonyms of face—the initial quantity of the progression, the first term : (that, from which as an origin the sequence commences.—Súr.)

Chaya, *prachaya* or *uttara*—the more (*adhika*—Súr.) or augment (*vriddhi*—Gang.) by which each term increases, the common difference. *Antya*, the last term. *Madhya*, the middle term. *Pada* or *gachchha*, the period, the number of terms : (so many days as the sequence reaches. —Súr.) *Sarvadhana*, *sredhi-phala* or *ganita*—the amount of the whole, the sum of the progression. It is called *ganita*, because it is found by computation (*gananá*).—Gan.

$= a+\left(\frac{n}{2}-1\right)b+\frac{1}{2}b=\frac{2a+(n-1)b}{2}$. And the sum of n terms $=\frac{n}{2}\left\{2a+(n-1)\,b\right\}$

Thus the rule holds good whether n be odd or even. The author does not notice that the first part of the rule in § 115 is only a particular case of the present rule.]

120. Example. A person, having given four *drammas* to priests on the first day, proceeded, my friend, to distribute daily alms at a rate increasing by five a day. Say quickly how many were given by him in half a month.

Statement : initial quantity 4 ; com. diff. 5 ; period 15.

Here, first term 4. Middle term 39. Last term 74. Sum 585.

121. Another example.[1] The initial term being seven, the increase five, and the period eight, tell me what the magnitudes of the middle and last terms are, and what the total sum is.

Statement : first term 7 ; com. diff. 5 ; period 8.

Answer : mean amount $\frac{49}{2}$. Last term 42. Sum 196.

Here, the period consisting of an even number of days, there is no middle day ; wherefore half the sum of the days preceding and following the mean place, must be taken for the mean amount : and the rule is thus proved.

[See note to § 119.]

[1] To exhibit an instance of an even number of terms, where there can consequently be no middle term (but a mean amount).—Gan.

122. Rule[1] : half a stanza. The sum of the progression being divided by the period, and half the common difference multiplied by one less than the number of terms, being subtracted, the remainder is the initial quantity.[2]

$$\left[s = \frac{n}{2}\left\{2a+(n-1)b\right\} = n\left\{a+\frac{n-1}{2}b\right\};\ \therefore\ \frac{s}{n}-\frac{n-1}{2}b=a,\right.$$

whence the rule.]

123. Example. We know the sum of the progression, one hundred and five ; the number of terms, seven ; the increase, three ; tell us, dear boy, the initial quantity.

Statement : com. diff. 3 ; period 7 ; sum 105.

Answer : first term, 6.

Rule[3] ; half a stanza.[4] The sum being divided by the period, and the first term subtracted from the quotient, the remainder, divided by half of one less than the number of terms, will be the common difference.[5]

$$\left[s=n\left\{a+\frac{n-1}{2}b\right\};\ \therefore\ \left\{\frac{s}{n}-a\right\}\div\frac{n-1}{2} = b,\right.$$ whence the rule.]

124. Example. On an expedition to seize his enemy's elephants, a king marched two *yojanas*

---

[1] The difference, period and sum being given, to find the first term.—Gan. and Súr.

[2] The rule is the converse of the preceding.—Gan. and Súr.

[3] The first term, period and sum being known, to find the common difference which is unknown.—Gan.

[4] Second half of one, the first half of which contained the preceding rule, § 122.

[5] This rule also is converse of the foregoing.—Gan.

the first day. Say, intelligent calculator, with what increasing rate of daily march he proceeded, he reaching his foe's city, a distance of eighty *yojanas*, in a week.

Statement : first term 2 ; period 7 ; sum 80.

Answer : com. diff. $\frac{22}{7}$.

125. Rule.[1] From the sum of the progression multiplied by twice the common increase, and added to the square of the difference between the first term and half that increase, the square root being extracted, this root less the first term and added to the (above-mentioned) portion of the increase, being divided by the increase, is pronounced[2] to be the period.

[The translation is rather obscure. A clearer rendering would be as follows:—"The sum of the progression multiplied by twice the common increase, being added to the square of the difference between the first term and half that increase, the square root of the result is extracted ; this root less, &c."

We have $n = \frac{n}{2}\left\{2a+(n-1)b\right\}$;

whence $n = \dfrac{b-2a \pm \sqrt{\{(2a-b)^2+8sb\}}}{2b}$

(Todhunter's Algebra, Art. 454)

$= \dfrac{1}{b}\left\{\dfrac{b}{2} - a \pm \sqrt{\left\{(a-\dfrac{b}{2})^2+2sb\right\}}\right.$, whence the

---

[1] The first term, common difference and sum being known, to find the period which is unknown.—Gan.

[2] By Brahmagupta and the rest.—Gan.

The rules are substantially the same ; the square being completed for the solution of the quadratic equation in the manner taught by Srídhara (cited in *Vija-ganita*, § 131) and by Brahmagupta.

rule, the *upper sign* only being taken by the author. He does not discuss the meaning of the *two* values of *n*.]

128. Example. A person gave three *drammas* on the first day, and continued to distribute alms increasing by two (a day); and he thus bestowed on the priests three hundred and sixty *drammas*: say quickly in how many days.

Statement: first term 3; com. diff. 2; sum 360.

Answer: period 18.

Geometrical Progression.

127. Rule[1]: a couplet and a half. The period being an uneven number, subtract one, and note 'multiplicator'; being an even one, halve it, and note 'square,' until the period be exhausted. Then the produce arising from multiplication and squaring (of the common multiplier) in the inverse order from the last,[2] being lessened by one, the remainder divided by the common multiplier less one, and multiplied by the initial quantity, will be the sum of a progression increasing by a common multiplier.[3]

---

[1] To find the sum of a progression, the increase being a multiplier.—Gan. In other words, to find the sum of an increasing geometrical progression.

[2] The last note is of course 'multiplicator.' For in exhausting the number of the period (when odd) you arrive at last at unity, an uneven number. The proposed multiplier (the common multiplicator of the progression) is therefore put in the last place; and the operations of squaring and multiplying by it are continued in the inverse order of the line of the notes.—Gan.

[3] The effect of squaring and multiplying, as directed, is the same as the continued multiplication of the multiplier for as many times as the num-

[Let a denote the first term and r the common ratio. Then, $a+ar+\ldots\ldots+ar^{n-1}=\frac{a(r^n-1)}{r-1}$. The first part of the rule is clumsily and obscurely stated. It is difficult to make out what the author means. He wants us to find r^n. Now if n is even and therefore of the form $2m$, we can find r^n by first squaring r, then squaring the square, and so on m times. If n is odd and therefore of the form $2m+1$, we can find r^n by first finding r^{2m} as before, and then multiplying the result by r. This is probably what the author means by the words multiplying and squaring. See the explanation of Ganesa in the foot-notes.]

128. Example. A person gave a mendicant a couple of cowry shells first, and promised a twofold increase of the alms daily. How many *nishkas* did he give in a month ?

Statement : first term, 2 ; increasing multiplier, 2 ; period 30.

Answer : 2147483646 cowries ; or 104857 *nishkas*, 9 *drammas*, 9 *panas*, 2 *kákinís*, and 6 shells.

129. Example. The initial quantity being two, my friend ; the daily augmentation, a threefold increase ; and the period, seven ; say what the sum in this case is.

---

ber of the period. For dividing by the multiplier the product of the multiplication continued to the uneven number, equals the product of multiplication continued to one less than the number ; and the extraction of the square root of a product of multiplication continued to the even number, equals continued multiplication to half that number. Conversely, squaring and multiplying equals multiplication for double and for one more time.—Gan.

Statement : first term, 2 ; increasing multiplier, 3 ; period, 7.

Answer : sum, 2186.

130—131. Rule :[1] a couplet and a half. The number of syllables in a verse being taken for the period, and the increase twofold, the produce of multiplication and squaring (as above directed, § 127) will be the number (of variations) of like verses.[2] Its square, and square's square, less their respective roots, will be (the variations) of alternately similar and of dissimilar verses (in tetrastics).[3]

[The rule refers specially to the example in § 132. It is a statement in a concrete shape of the following proposi-

1 Incidentally introduced in this place, showing a computation serviceable in prosody.—Súr and *Mano*. To calculate the variations of verse, which are also found by the sum of permutations (§113).—Gan.

2 Sanskrit prosody distinguishes metre in which the four verses of the stanza are alike, or the alternate ones only so, or all four dissimilar. Asiat. Res., Vol. X, syn. tab., v, vi and vii.

3 The number of possible varieties of verse found by the rule of permutation (§113) is the same as the continued multiplication of two : this number being taken, because the varieties of syllables are so many, long and short. Accordingly this is assumed for the common multiplier. The product of its continued multiplication is to be found by this method of squaring and multiplying (§127) ; assuming for the period a number equal to that of syllables in the verse. The varieties of similar verses are the same as those of one verse containing twice as many syllables ; and the changes in the four verses are the same as those of one verse comprising four times as many syllables, excepting, however, that these permutations embracing all the possible varieties, comprehend those of like and half-alike metre. Wherefore the number first found is squared, and this again squared, corresponding to twice or four times the number of places ; and the roots of these squares are subtracted [for obtaining the varieties of alternately like and dissimilar verses respectively,—Ed.]—Gan.

tion :—The total number of combinations of n things taken, 1, 2, ... n at a time $= 2^n - 1$. (See note to § 113).]

132. Example. Tell me directly the number (of varieties) of like, alternately like, and dissimilar verses respectively, in the metre named *anushtubh*.[1]

Statement : increasing multiplier, 2 ; period, 8.

Answer : variations of like verses, 256 ; of alternately alike verses, 65280 ; of dissimilar verses, 4294901760.

[The total number of syllables in the four *charanas* of the *anushtubh* metre being 32 (8 to each *charana*), the possible varieties of arrangements of long and short syllables in the metre are $2^{32} = 4294967296$. (See note to § 113). These evidently include cases of (*a*) all like, and (*b*) alternately like *charanas*. To find the number of these cases, we find the number of varieties of the syllables in *two charanas* ; which number is 2^{16} or 65536. It is clear that if we place each one of these varieties under itself, we shall get all the cases included in (*a*) and (*b*). Hence the total number of cases in (*a*) and (*b*) is 65536. Of these the number of cases in (*a*) clearly is the number of varieties that may occur in one *charana* $= 2^8 = 256$. Consequently the number of cases in (*b*) is 65536 − 256 or 65280. Subtracting 65536 from 4294967296 we get 4294901760, which is considered as the number of *dissimilar* verses or *charanas*. It is to be observed, however, that these last include cases in which the first two *charanas* are like, as also the last two ; or cases in which the first two are like, but the last two unlike ; and so forth. These cases are not separately considered.]

---

[1] Asiat. Res., Vol. X, p. 438 ; syn. tab., p. 469.

CHAPTER VI.

PLANE FIGURE.[1]

133. Rule. A side[2] is assumed. The other side

---

[1] *Kshetra-vyavahára*, determination of plane figure. *Kshetra*, as expounded by Ganesa, signifies plane surface bounded by lines, straight or curved; as triangle, &c. *Vyavahára* is the ascertainment of its dimensions, as diagonal, perpendicular, area, &c. Ganesa says plane figure is fourfold; triangle, quadrangle, circle and bow. Triangle (*tryasra, trikona* or *tribhuja*) is a figure containing (*tri*) three (*asra* or *kona*) angles, and consisting of as many (*bhuja*) sides. Quadrangle or tetragon (*chaturasra, chatushkona, chaturbhuja*), is a figure comprising (*chatur*) four (*asra*, &c.) angles or sides. The circle and bow, he observes, need no definition. Triangle is either (*játya*) right-angled, as that which is first treated of in the text; or it is (*tribhuja*) trilateral (and oblique) like the fruit of the *sringáta* (Trapa natans). This again is distinguished according as the (*lamba*) perpendicular falls within or without the figure: *viz*., *antarlamba*, acute-angled; *bahirlamba*, obtuse-angled. Quadrangle also is in the first place twofold: with equal, or with unequal, diagonals. [This is not a proper classification.—Ed.] The first of these, or equidiagonal tetragon (*sama-karna*), comprises four distinctions: 1st, *sama-chaturbhuja*, equilateral, a square; 2d. *vishama-chaturbhuja*, a trapezium; 3d. *áyata-dírgha-chaturasra*, an oblique parallelogram; [this is not correct; for a parallelogram with equal diagonals must be either a rectangle or a square, so that this 3d. cannot be a distinct species.—Ed.); 4th. *áyata-samalamba*, oblong with equal perpendiculars, *i.e.*, a rectangle. The second sort of quadrangle, or the tetragon with unequal diagonals (*vishamakarna*), embraces six sorts: 1st. *sama-chaturbhuja*, equilateral, a rhombus; 2d. *sama-tribhuja*, having three equal sides; 3d. *sama-dwi-dwi-bhuja*, consisting of two pairs of equal sides, a rhomboid; 4th. *sama-dwi-bhuja*, having two equal sides; 5th. *vishama-chaturbhuja*, composed of four unequal sides, a trapezium; 6th. *sama-lamba*, having equal perpendiculars, a trapezoid. The several sorts of figures, observes the commentator, are fourteen; the circle and bow being but of one kind each. He adds, that pentagons (*panchásra*), &c. comprise triangles (and are reducible to them).

[2] *Báhu*, *dosh*, *bhuja*, and other synonyms of arm are used for the leg of a triangle, or side of a quadrangle or polygon; so called, as resembling the human arm.—Gan. and Súr.

in the rival direction is called the upright,[1] whether in a triangle or tetragon, by persons conversant with the subject.

134. The square root of the sum of the squares of those legs is the diagonal.[2] The square root, extracted from the difference of the squares of the diagonal and side, is the upright; and that, extracted from the difference of the squares of the diagonal and upright, is the side.[3]
[Euclid I. 47.]

135.[4] Twice the product of two quantities, added to the square of their difference, will be the sum of their squares. The product of their sum and

---

[1] Either leg being selected to retain this appellation, the others are distinguished by different names. That which proceeds in the opposite direction, meaning at right angles, is called *koti*, *uchchhráya*, *uchchhriti*, or any other term signifying upright or elevated. Both are alike sides of the triangle or of the tetragon, differing only in assumed situation and name.—Gan. and Súr.

[2] A thread or oblique line from the two extremities of the legs, joining them, is the *karna*, also termed *sruti*, *sravana*, on any other word signifying ear. It is the diagonal of a tetragon.—Súr., Rang., &c. Or, in the case of a triangle, it is the diagonal of the parallelogram, whereof the triangle is the half: and is the hypotenuse of a right-angled triangle.

[3] The rule concerns (*játya*) right-angled triangles. The proof is given both algebraically and geometrically by Ganesa (*upapatti avyakta-kriyayá*, proof by algebra; *kshetragatopapatti*, geometrical demonstration); and the algebraical proof is also given by Súryadása. Ranganátha cites one of those demonstrations from his brother Kamalákara, and the other from his father Nrisinha, in the *Várttika*, or critical remarks on the (*Vásaná*) annotations of the *Siromani*; and censures the *Sringára-tilaka* for denying any proof of the rule besides experience. Bháskara has himself given a demonstration of the rule in his *Víja-ganita*, § 146.

[4] A stanza of six *charanas* of *anushtubh metre*.

difference will be the difference of their squares : as must be everywhere understood by the intelligent calculator.[1]

[$2ab+(a-b)^2=a^2+b^2$.

$(a+b)\ (a-b)=a^2-b^2$.

Geometrical proofs of these formulæ are furnished by Euc. II. 5 and 9. The object of introducing them here is to facilitate the calculations required in § 134.]

136. Example. Where the upright is four and the side three, what is the hypotenuse ? Tell me also the upright from the hypotenuse and side ; and the side from the upright and hypotenuse.

Statement : side 3 ; upright 4.

Sum of their squares 25. Or, the product of the sides, doubled, 24 ; square of the difference, 1 ; added together, 25. The square root of this is the hypotenuse 5.

Difference of the squares 5 and 3 is 16. Or the sum 8 multiplied by the difference 2, makes 16. Its square root is the upright 4.

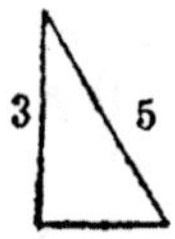

Difference of squares, found as before, 9. Its square root is the side 3.

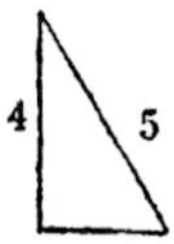

---

[1] Ganesa here also gives both an algebraic and a geometrical proof of the latter rule ; and an algebraical one only of the first. See *Vija-ganita* under § 148, whence the latter demonstration is borrowed ; and § 147, where the first of the rules is given and demonstrated.

137. Example. Where the side measures three and a quarter, and the upright, as much ; tell me quickly, mathematician, what the length of the hypotenuse is.

Statement : side $\frac{13}{4}$; upright $\frac{13}{4}$. Sum of the squares $\frac{338}{16}$ or $\frac{169}{8}$. Since this has no (assignable) root, the hypotenuse is a surd. A method of finding its approximate root follows :—

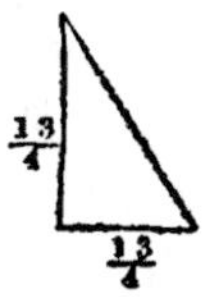

138. Rule. From the product of numerator and denominator,[1] multiplied by any large square number assumed, extract the square root : that, divided by the denominator taken into the root of the multiplier, will be an approximation.

The square of the above hypotenuse, $\frac{169}{8}$ is proposed). The product of its numerator and denominator is 1352. Multiplied by a myriad (the square of a hundred) the product is 13520000. Its root is 3677 nearly.[2] This divided by the denominator taken into the square root of the multiplier, *viz.*, 800, gives the approximate root $4\frac{477}{800}$. It is the hypotenuse. So in every similar instance.

$$\left[\sqrt{\frac{a}{b}}=\frac{\sqrt{a\times b}}{b}=\frac{\sqrt{a\times b\times c^2}}{bc}.\right.$$

The object of multiplying the product of numerator and denominator by a large square number (being some power of ten, as the above process shows), and then

[1] If the surd be not a fraction, unity may be put for the denominator, and the rule holds good.—Gan.

[2] The remainder being unnoticed.

taking the square root approximately is practically to get the square root to a certain number of decimal places. Bháskara, however, does not use the decimal notation which was probably not known in his time, and expresses the result as a fraction. In the above example, the result obtained will be found to be correct to two decimal places.]

139. Rule.[1] A side is put. From the multiplied by twice some assumed number, and divided by one less than the square of the assumed number, an upright is obtained. This, being set apart, is multiplied by the arbitrary number, and the side as put is subtracted; the remainder will be the hypotenuse. Such a triangle is termed right-angled.[2]

[Let a denote the given side, and n the assumed number.

Then proceeding by the rule, we get $\frac{2an}{n^2-1}$ for upright, and $\frac{2an}{n^2-1} \times n - a = a\frac{n^2+1}{n^2-1}$ for hypotenuse. To verify this we have $a^2 + \left(\frac{2an}{n^2-1}\right)^2 = \frac{a^2}{(n^2-1)^2}\left\{ (n^2-1)^2 + 4n^2 \right\}$ $= \frac{a^2(n^2+1)^2}{(n^2-1)^2} = \left(a\frac{n^2+1}{n^2-1}\right)^2$. The proof of this rule given by Súryadása shows how these expressions for the upright and hypotenuse are arrived at, although it is rather difficult to follow it. The quantities $2n$, n^2-1,

---

[1] Either the side or upright being given, to find the other two sides.—Súr. To find the upright and hypotenuse from the side; or the side and hypotenuse from the upright.—Gaṇ. The problem is an indeterminate one, as is intimated by the author.

[[2] Colebrooke uses the word *rectangular*. But the more usual word in modern geometry is *right-angled*.—Ed.]

n^2+1 may be taken to represent the upright, side and hypotenuse of a right-angled triangle, because $(2n)^2+(n^2-1)^2=(n^2+1)^2$. Now consider another right-angled triangle similar to the above, the side being a. Then, since the sides of the two triangles are proportional, the upright of the second triangle will obviously be $\frac{2an}{n^2-1}$.

Again, as $n\times$upright of the first triangle = its side + its hypotenuse, so $n\times$upright of the second triangle = its side + its hypotenuse. Thus, $n\times\frac{2an}{n^2-1} = a+$hypotenuse, whence hypotenuse $= n\times\frac{2an}{n^2-1} - a$. Thus we see how the expressions are got.]

140. Or a side is put. Its square, divided by an arbitrary number, is set down in two places: and the arbitrary number being added and subtracted, and the sum and difference halved, the results are the hypotenuse and upright.[1] Or, in like manner, the side and hypotenuse may be deduced from the upright. Both results are rational quantities.

[Let a denote the given side, and n the assumed number

Then by the rule we have $\frac{1}{2}\left(\frac{a^2}{n}+n\right)$ for hypotenuse,

---

[1] Assume any number for the difference between the upright and hypotenuse. The difference of their squares (which is equal to the square of the given side) being divided by that assumed difference, the quotient is the sum of the upright and hypotenuse. For the difference of the squares is equal to the product of the sum and difference of the roots (§ 135). The upright and hypotenuse are therefore found by the rule of concurrence (§ 55).—Gan.

and $\frac{1}{2}\left(\frac{a^2}{n} - n\right)$ for upright. To verify this we have

$$a^2 + \frac{1}{4}\,\frac{(a^2 - n^2)^2}{n^2} = \frac{4a^2n^2 + (a^2 - n^2)^2}{4n^2} = \frac{(a^2 + n^2)^2}{4n^2} = \left\{\frac{1}{2}\left(\frac{a^2}{n} + n\right)\right\}$$

Ganesa gives an elegant demonstration of this rule, based on the fact that the assumed number n is the difference between the hypotenuse and upright, as is obviously the case. See foot-note.]

141. Example: The side being in both cases twelve, tell quickly by both methods, several uprights and hypotenuses, which shall be rational numbers.

Statement : side 12 ; assumption 2. The side, multiplied by twice that, *viz.*, 4, is 48. Divide by the square of the arbitrary number less one, *viz.*, 3, the quotient is the upright 16. This upright multiplied by the assumed number is 32, from which subtract the given side ; the remainder is the hypotenuse 20.

Assume 3. The upright is 9, and the hypotenuse 15. Or, putting 5, the upright is 5, and the hypotenuse 13.

By the second method : the side, as put, 12. Its square 144. Divide by 2, the arbitrary number being 2, the quotient is 72. Add and subtract the arbitrary number, and halve the sum and difference. The hypotenuse and upright are found, *viz.*, hypotenuse 37, upright 35.

Assume 4. The upright is 16, and the hypotenuse 20. Assuming 6, the upright is 9, and the hypotenuse 15.[1]

142. Rule.[2] Twice the hypotenuse taken into an arbitrary number, boing divided by the square of the arbitrary number added to one, the quotient is the upright. This taken apart is to be multiplied by the number put : the difference between the product and the hypotenuse is the side.

[Let a denote the hypotenuse, and n the assumed number Then, by the rule, the upright is $\frac{2an}{n^2+1}$, and the side, $\frac{2an^2}{n^2+1}-a=a\frac{n^2-1}{n^2+1}$. For $\left(\frac{2an}{n^2+1}\right)^2+\left(a\frac{n^2-1}{n^2+1}\right)^2=a^2$. The proof of this rule given by Súryadása is exactly similar to that of the rule in §139. See note to §139]

143. Example. The hypotenuse being measured by eighty-five, say promptly, learned man, what uprights and sides will be rational.

Statement : hypotenuse 85. This doubled is 170, and multiplied by an arbitrary number two is 340. This, divided by the square of the arbitrary number added to one, *viz.*, 5, is the upright 68. This upright multiplied by the arbitrary number makes 136 ; and subtracting the hypotenuse, the

---

[1] In like manner, if the upright be given 16, its square 256 divided by the arbitrary number 2 is 128. The arbitrary number, subtracted and added, makes 126 and 130 ; which halved give the side 63, and the hypotenuse 65.—Gang. and Súr.

[2] From the hypotenuse given, to find the side and upright in rational numbers.—Gan. The problem is an indeterminate one.

side comes out 51. Or putting four, the upright will be 40, and the side 75.

144. Rule. Or else the hypotenuse is doubled and divided by the square of an assumed number added to one. The hypotenuse less that quotient is the upright. The same quotient multiplied by the assumed number is the side.[1]

The same hypotenuse 85. Putting two, the upright and side are 51 and 68. Or, with four, they are 75 and 40.

Here the distinction between side and upright is in name only, and not essential.

[Taking a and n as in § 142, we get $a-\frac{2a}{n^2+1}=a\,\frac{n^2-1}{n^2+1}$ for upright, and $\frac{2an}{n^2+1}$ for side. Verification as in § 142. To see how these expressions are arrived at, take $2n$, n^2-1 and n^2+1 for the side, upright and hypotenuse of a right-angled triangle, and proceed as in § 139, note.]

145. Rule.[2] Let twice the product of two assumed numbers be the upright; and the difference of their squares, the side: the sum of their squares will be the hypotenuse, and a rational number.

---

[1] This and the preceding rule are founded on the same principle, differing only in the order of the operation and names of the sides: the same numbers come out for the side and upright in one mode, which were found for the upright and side by the other.

[2] Having taught the mode of finding a third side from any two of hypotenuse, upright and side; and in like manner from one, the other two; the author now shows a method of finding all three rational (none being given).—Gan. The problem is an indeterminate one.

[Let a and b be the assumed numbers.

Then $2ab$ is the upright, and a^2-b^2 the side. The hypotenuse is $\sqrt{(2ab)^2+(a^2-b^2)^2}=a^2+b^2$.

Thus the three sides are all rational. Ganesa gives a proof of this rule after the manner of the *Vija-ganita*; but it is very obscure and cannot be easily followed.]

146. Example. Tell quickly, friend, three numbers, none being given, with which as upright, side and hypotenuse, a right-angled traingle may be (constructed). Let two numbers be put, 1 and 2. From these, the side, upright and hypotenuse are found, 4, 3, 5. Or, putting 2 and 3, the side, upright and hypotenuse deduced from them are, 12, 5, 13. Or let the assumed numbers be 2 and 4 : from which will result 16, 12, 20. In like manner, manifold (answers are obtained).

147. Rule.[1] The square of the ground intercepted between the root and tip is divided by the (length of the) bambu, and the quotient severally added to, and subtracted from, the bambu : the moieties (of the sum and difference) will be the two portions of it representing hypotenuse and upright.[2]

---

[1] The sum of hypotenuse and upright being known, as also the side, to discriminate the hypotenuse and upright.—Gan.

[2] The height from the root to the fracture is the upright. The remaining portion of the bambu is the hypotenuse. The whole bambu, therefore, is the sum of hypotenuse and upright. The ground intercepted between the root and tip is the side ; it is equal to the square root of the difference between the squares of the hypotenuse and upright, Hence the square of the side, divided by the sum of hypotenuse and upright, is their difference (§ 135). With these (sum and difference) the upright and hypotenuse are found by the rule of concurrence (§ 55).—Gan.

[The rule bears reference to the example which follows.

Let a denote the height of the bambu, b, the distance between root and tip, and x the height at which the bambu is broken.

Then, $b^2=(a-x)^2-x^2$

$$\therefore \frac{b^2}{(a-x)+x}=(a-x)-x$$

$$i.e. \frac{b^2}{a}=(a-x)-x$$

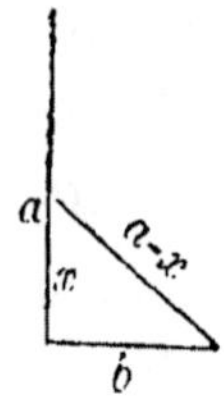

also $a=(a-x)+x$, identically.

Hence $\frac{1}{2}\left(a+\frac{b^2}{a}\right)=a-x$ or hypotenuse, and $\frac{1}{2}\left(a-\frac{b^2}{a}\right)=x$ or upright, whence the rule. The above proof is the same as that given by Ganesa. See foot-note.]

148. Example. If a bambu, measuring thirty-two cubits and standing upon level ground, be broken in one place by the force of the wind, and the tip of it meet the ground at sixteen cubits: say, mathematician, at how many cubits from the root it is broken.

Statement. Bambu 32. Interval between the root and tip of the bambu, 16. It is the side of the triangle. Proceeding as directed, the upper and lower portions of the bambu are found to be 20 and 12.

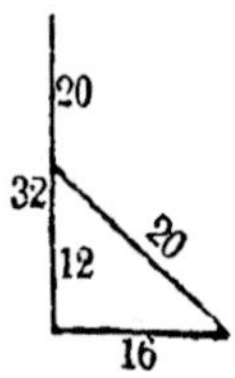

149. Rule.[1] The square (of the height) of

---

[1] The sum of the side and hypotenuse being known, as also the upright, to discriminate the hypotenuse and side.—Gan. The rule bears reference to the example which follows. The principle is the same as that of the preceding rule.

the pillar is divided by the distance of the snake from his hole ; the quotient is to be subtracted from that distance. The meeting of the snake and peacock is from the snake's hole half the remainder, in cubits.

150. Example. A snake's hole is at the foot of a pillar, nine cubits high, and a peacock is perched on its summit. Seeing a snake at the distance of thrice the pillar gliding towards his hole, he pounces obliquely upon him. Say quickly at how many cubits from the snake's hole they meet, both proceeding an equal distance.

Statement. Pillar 9. It is the upright. Distance of the snake from his hole, 27. It is the sum of hypotenuse and side.

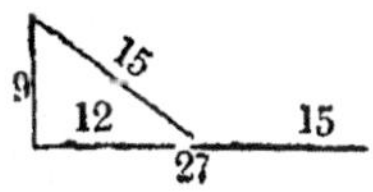

Proceeding as directed, the distance between the hole and the place of meeting is found to be 12 cubits.[1]

[The principle of the rule in § 149 is, as Colebrooke observes, and as is also evident from the example in § 150, the same as that of the rule in § 147. The peacock is supposed not to change his direction, and to pounce in such a direction that the distance traversed by him being the hypotenuse of a right-angled triangle, is equal to the distance traversed by the snake. Practically, however, such a thing does not happen ; but the bird of prey

---

1. Subtracted from the sum of hypotenuse and side, this leaves 15 for the hypotenuse. The snake had proceeded the same distance of 15 cubit towards his hole, as the peacock in pouncing upon him. Their progress is therefore equal.—Súr.

changes its direction at every instant, and describes a curved path known as the curve of pursuit. See Tait and Steele's Dynamics of a Particle, Art. 33.

Let a denote the distance of the snake from the hole, b the height of the pillar, and x the distance required.

Then, $b^2=(a-x)^2-x^2$, whence as in § 147, $x=\frac{1}{2}\left(a-\frac{b^2}{a}\right)$. Hence the rule.]

151. Rule.[1] The quotient of the square of the side divided by the difference between the hypotenuse and upright is twice set down; and the difference is subtracted from the quotient (in one place) and added to it (in the other). The moieties (of the remainder and sum) are in their order the upright and hypotenuse.[2]

This[3] is to be generally applied by the intelligent mathematician.

[The demonstration of the rule on § 147 applies to this rule as Ganesa observes.

Let a denote the difference between hypotenuse and upright, b the side, and x the upright.

Then, $b^2=(a+x)^2-x^2$, whence as in § 147, $x=\frac{1}{2}\left(\frac{b^2}{a}-a\right)$, and $a+x=\frac{1}{2}\left(\frac{b^2}{a}+a\right)$. Hence the rule.]

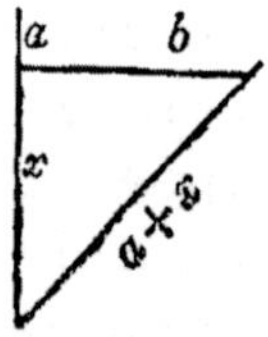

[1] The difference between the hypotenuse and upright being known, as also the side, to find the upright and hypotenuse.—Gan.

[2] The demonstration, distinctly set forth under a preceding rule, is applicable to this.—Gan.

[3] Beginning from the instance of the broken bambu (§ 147) and including what follows.—Gan.

152. Friend, the space between the lotus (as it stood) and the spot where it submerged, is the side. The lotus as seen (above water) is the difference between the hypotenuse and upright. The stalk is the upright, for the depth of water is measured by it. Say, what the depth of the water is.[1]

153. Example.[2] In a certain lake swarming with ruddy geese[3] and cranes, the tip of a bud of lotus was seen a span above the surface of the water. Forced by the wind it gradually advanced, and was submerged at the distance of two cubits. Compute quickly, mathematician, the depth of the water.

Statement. Diff. of hypotenuse and upright, $\frac{1}{2}$ cubit. Side 2 cubits. Proceeding as directed, the upright is found $\frac{15}{4}$.

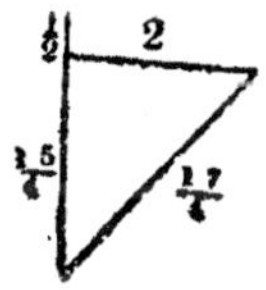

It is the depth of the water. Adding to it the height of the bud, the hypotenuse comes out $\frac{17}{4}$.

[Let O be the root of the lotus, A its tip, and C the point on the surface of the water where it is submerged. Then, while it advances by the force of the wind, O remains fixed, and the lotus describes an arc of a circle, of which O is the centre, and OA the

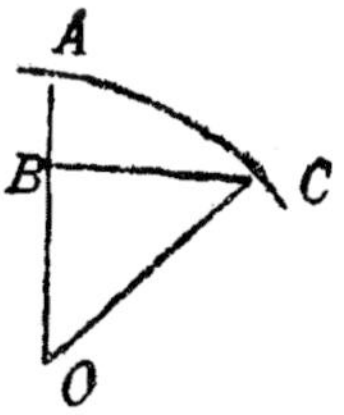

[1] The sides constituting the figure in the example which follows, are here set forth, to assist the apprehension of the student.—Súr. and Gan.

[2] [This example is inserted in Barnard Smith's Arithmetic, Appendix, p. 300.—Ed.]

[3] Anas Casarca.

radius. Hence $OC=OA$. Then, as the author himself explains in § 152, the solution will follow from the method of § 151, where we have only to substitute $\frac{1}{2}$ for a, and 2 for b.]

154. Rule.[1] The height of the tree multiplied by its distance from the pond, is divided by twice the height of the tree added to the space between the tree and the pond : the quotient will be the measure of the leap.

[The rule refers to the example which follows.

Let D be the top of the tree, and B the position of the pond. The first ape is supposed to descend from D to A, and then to go from A to B ; while the second ape is supposed to jump vertically upwards from D to C, and then to leap directly from C to B. Now let $DA=a$, $AB=b$, and $CD=x$, which is required. Then by the question, we have $x+\sqrt{(a+x)^2+b^2}=a+b$;

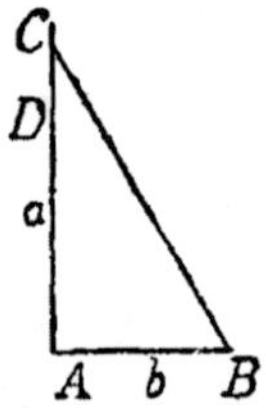

$$\therefore\ (a+x)^2+b^2=(a+b)^2-2(a+b)x+x^2,$$

whence $x=\dfrac{ab}{2a+b}$. Hence the rule.]

155. Example. From a tree a hundred cubits high, an ape descended and went to a pond two hundred cubits distant : while another ape, vaulting to some height off the tree, proceeded with velocity diagonally to the same spot. If the

[1] The sum of the hypotenuse and upper portion of the upright being given, and the lower portion being known, as also the side : to discriminate the upper portion of the upright from the hypotenuse.—Gan. As in several preceding instances, a reference to the example is requisite to the understanding of the rule.

space travelled by them be equal, tell me quickly, learned man, the height of the leap, if thou have diligently studied calculation.

Statement. Tree 100 cubits. Distance of it from the pond 200. Proceeding as directed, the height of the leap comes out 50.

156. Rule.[1] From twice the square of the hypotenuse subtract the sum of the upright and the side multiplied by itself, and extract the square root of the remainder. Set down the sum twice, and let the root be subtracted in one place and added in the other. The moieties will be measures of the side and upright.[2]

[Let a denote the sum of side and upright, and b the hypotenuse. Also let x denote the upright. Then we have evidently, $(a-x)^2+x^2=b^2$

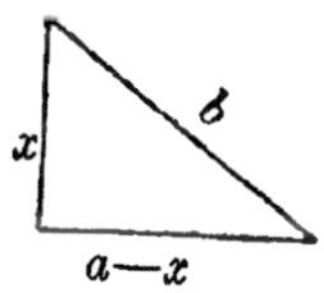

---

[1] Hypotenuse being known, as also the sum of the side and upright, or their difference; to discriminate those sides.—Gan.

[2] In like manner, the difference of the side and upright being given, the same rule is applicable.—Gan. [A slight variation will be necessary; see note to § 158.—Ed.] The principle of the rule is this: the square of the hypotenuse is the sum of the squares of the sides. But the sum of the squares with twice the product of the sides added to it is the square of the sum; and with the same subtracted is the square of the difference. Hence cancelling equal quantities affirmative and negative, twice the square of the hypetonuse will be the sum of the squares of the sum and difference. Therefore, subtracting from twice the square of hypotenuse the square of the sum, the remainder is the square of the difference; or conversely, subtracting the square of the difference, the residue is the square of the sum. The square root is the sum or difference. With these, the sides are found by the rule of concurrence.—Gan. and Súr.

$\therefore\ 2x^2-2ax+a^2\quad b^2=0,$

whence $x=\dfrac{a\pm\sqrt{2b^2-a^2}}{2}$.

If we take the upper sign, then $a-x=\dfrac{a-\sqrt{2b^2-a^2}}{2}$, and if we take the lower sign, then $a-x=\dfrac{a+\sqrt{2b^2-a^2}}{2}$. It is evident that we may take either sign. The reason for the rule is obvious. The method of solving an adfected quadratic equation by completing the square has been mentioned before, See § § 62—63. Interesting geometrical interpretations of the expressions for x and $a-x$ are given by Ganesa and Súryadása. See foot-note.]

157. Example. Where the hypotenuse is seven above ten; and the sum of the side and upright, three above twenty; tell them to me, my friend.

Statement: Hypotenuse 17; sum of side and upright 23. Proceeding as directed, the side and upright are found 8 and 15.

158. Example.[1] Where the difference of the side and upright is seven, and hypotenuse is thirteen, say quickly, eminent mathematician, what the side and upright are.

Statement: hypotenuse 13; difference of side and upright 7. Proceeding as directed, the side and upright come out 5 and 12.

---

[1] This example of a case where the difference of the sides is given, is omitted by Súryadása but noticed by Ganesa. Copies of the text vary: some containing, and others omitting, the instance.

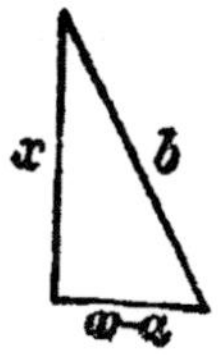

[Let x denote the upright, and a the diff. between upright and side, the upright being supposed > side. Then $x-a$ will denote the side; and we evidently have $x^2+(x-a)^2=b^2$, whence as in § 156, $x=\frac{a\pm\sqrt{2b^2-a^2}}{2}$. But here we must take the upper sign alone, since $x-a$ is necessarily positive. Thus we have $x=\frac{\sqrt{2b^2-a^2}+a}{2}$, and $x-a=\frac{\sqrt{2b^2-a^2}-a}{2}$. From these values it is clear that the rule in § 156 must be slightly varied in order to be applicable to the present case.]

159. Rule.[1] The product of two erect bambus being divided by their sum, the quotient is the perpendicular[2] from the junction (intersection) of threads passing reciprocally from the root (of one) to the tip (of the other). The two bambus, multiplied by an assumed base, and divided by their sum, are the portions of the base on the respective sides of the perpendicular.

[From similar triangles (see figure) we have

$$\frac{p}{a}=\frac{y}{x+y},$$

$$\frac{p}{b}=\frac{x}{x+y};$$

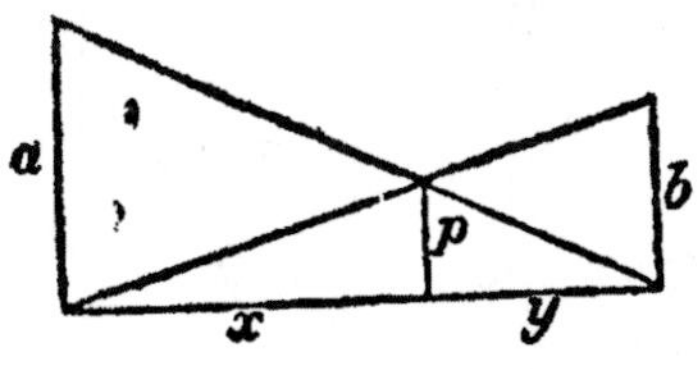

[1] Having taught fully the method of finding the sides in a right-angled triangle, the author next propounds a special problem.—Gan. To find the perpendicular the base being unknown.—Súr.

[2] *Lamba, avalamba, valamba, adholamba*, the perpendicular.

$$\therefore p\left(\frac{1}{a}+\frac{1}{b}\right)=1, \text{ and } \therefore p=\frac{ab}{a+b}.$$

Thus p is independent of x and y, provided a and b be given.

Again, let $x+y=k$, any assumed number.

Then $x=\frac{pk}{b}=\frac{ak}{a+b}$,

and $y=\frac{pk}{a}=\frac{bk}{a+b}$.

This rule shows that the property of similar triangles was known. See Bháskara's remark at the end of § 160, and the proof given by Ganesa, cited in the foot-note, which is slightly different and more cumbrous.]

160. Example. Tell the perpendicular drawn from the intersection of strings stretched mutually from the roots to the summits of two bambus fifteen and ten cubits high, standing upon ground of unknown extent.

Statement : bambus 15, 10. The perpendicular is found 6.

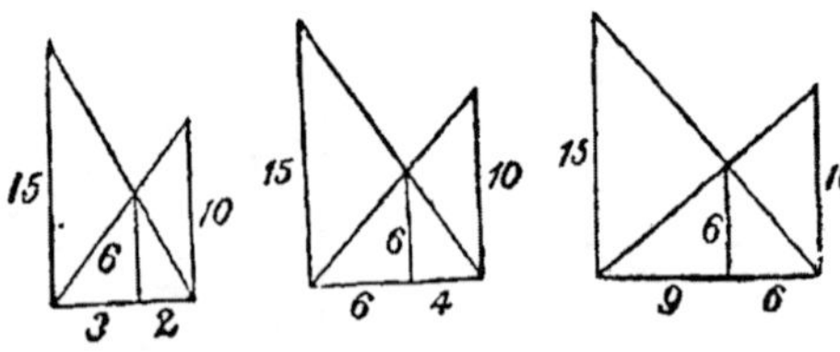

Next to find the segments of the base. Let the ground be assumed 5; the segments come out 3 and 2. Or putting 10, they are 6 and 4. Or taking 15, they are 9 and 6.

See the figures. In every instance the perpendicular is the same,[1] *viz.*, 6.

The proof is in every case by the rule of three: if with a side equal to the base, the bambu be the upright, then with the segment of the base, what will be the upright ?[2]

161. Aphorism.[3] That figure, though rectilinear, of which sides are proposed by some presumptuous person, wherein one side[4] exceeds or equals the sum of the other sides, may be known to be no figure.

---

[1] However the base may vary by assuming a greater or less quantity for it, the perpendicular will always be the same.—Gan.

[2] On each side of the perpendicular, are segments of the base relative to the greater and smaller bambus, and larger or less analogously to them. Hence this proportion : "If with the sum of the bambus, this sum of the segments equal to the entire base be obtained, then, with the smaller bambu, what is had ?" [This proportion cannot be at once obtained easily, but may be got by dividing corresponding members of the first two equations in the note §159, whence we have $\frac{a}{b}=\frac{x}{y}$, and therefore $\frac{a+b}{x+y}=\frac{b}{y}$.—Ed.] The answer gives the segment which is relative to the least bambu. Again : "If with a side equal to the whole base, the higher bambu be the upright, then with a side equal to the segment found as above, what is had ?" The answer gives the perpendicular let fall from the intersection of the threads. Here a multiplicator and a divisor equal to the entire base are both cancelled as equal and contrary : and there remain the product of the two bambus for numerator and their sum for denominator. Hence the rule.—Gan.

[3] The aphorism explains the nature of impossible figures proposed by dunces.—Súr. It serves as a definition of plane figure (*kshetra*).—Gan. In a triangle or other plane rectilinear figure, one side is always less than the sum of the rest. If equal, the perpendicular is nought, and there is no complete figure. If greater, the sides do not meet.—Súr. Containing no area, it is no figure.—*Kaumudí* cited by Ranganátha.

[4] The principal or greatest side.—Gan, *Kaum*, Rang.

[This follows from Euclid I. 20. Any side of a triangle or of any polygon must be less than the sum of the remaining sides. Hence the numbers 2, 3, 6, 12 cannot represent the sides of a quadrilateral.]

162. Example. Where sides are proposed two, three, six and twelve in a quadrilateral, or three, six and nine in a triangle, by some presumptuous dunce, know it to be no figure.

Statement. The figures are both incongruous. Let straight rods exactly of the lengths of the proposed sides be placed on the ground, the incongruity will be apparent.[1]

163—164. Rule[2] in two couplets. In a triangle, the sum of two sides being multiplied by their difference, is divided by the base[3]; the quotient is subtracted from, and added to, the base which is

---

[1] The rods will not meet.—Súr.

[2] In any triangle to find the perpendicular, segments and area. This is introductory to a fuller consideration of areas. Gan. and Súr.

[3] *Bhúmi, bhú, ku, mahí* or any other term signifying earth; the ground or base of a triangle or other plane figure. Any one of the sides is taken for the base, and the rest are termed simply sides. Ganesa restricts the term to the greatest side. See note § 168.

Lamba, &c., the perpendicular. See note § 159. *Abádhá, abadhá, avabádhá,* segment of the base made by the perpendicular. These are terms introduced by earlier writers. These segments are internal in an acute-angled triangle, but external in an obtuse-angled one. *Phala, ganita, kshetra-phala, sama-koshtha-miti:* the measure of like compartments, or number of equal squares of the same denomination (as cubit, fathom, finger, &c.) in which the dimension of the side is given ; the area or superficial content.—Gan. and Súr.

twice set down : and being halved, the results are segments corresponding to those sides.[1]

The square root of the difference of the squares of the side and its own segment of the base becomes the perpendicular. Half the base multiplied by the perpendicular[2] is in a triangle the exact[3] area.[4]

[$x+y$ is given, as also a and b.

We have $a^2-b^2=x^2-y^2$,

$$\frac{a^2-b^2}{x^2-y^2}=x-y.$$

Then $(x+y)+(x-y)=2x$.

and $(x+y)-(x-y)=2y$;

and half of these results give x and y.

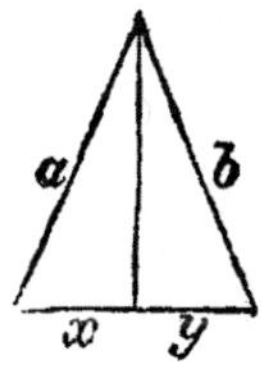

Also perpendicular $=\sqrt{a^2-x^2}$,

and area $=\frac{1}{2}\times \text{base}\times \text{altitude}$.

---

[1] The relative or corresponding segments. The smaller segment answers to the less side, and the larger to the greater side.—Gan.

[2] Or half the perpendicular taken into the base.—Gan.

[3] *Sphuta-phala*, distinct or precise area ; opposed to *asphuta*—or *sthūla-phala*, indistinct or gross area. See § 167.

[4] Demonstration. In both the right-angled triangles formed in the proposed triangle, one on each side of the perpendicular, this line is the upright ; the side is hypotenuse, and the corresponding segment is side. Hence, subtracting the square of the perpendicular from the square of the side, the remainder is the square of the segment. So, subtracting the square of the other side, there remains the square of the segment answering to it. Their difference is the difference of the squares of the segments, and is equal to the difference of the squares of the sides, since an equal quantity has been taken from each ; for any two quantities less an equal quantity have the same difference. It is equal to the product of the sum and difference of the simple quantities. Therefore, the sum of the sides multiplied by their difference is the difference of the squares of the segments. But the base is the sum of the segments. The difference of

The above is practically the demonstration given by Ganesa, although it is rather long as it is expressed in words.]

165. Example. In a triangular figure in which the base is fourteen and its sides thirteen and fifteen, tell quickly the length of the perpendicular, the segments, and the dimension by like compartments termed area.

Statement base 14 ; sides 13 and 15. Proceeding as directed the segments, are found 9 and 5, the perpendicular, 12 ; the area 84.

166. Example. In a triangle, wherein the sides measure ten and seventeen, and the base nine, tell me promptly, expert mathematician, the segments, perpendicular, and area.

---

the squares, divided by that, is the difference of the segments. From which by the rule of concurrence (§ 55) the segments are found.

The square root of the difference between the squares of the side and segment (taken as hypotenuse and side) is the upright or perpendicular.

Dividing the triangle by a line across the middle (of the perpendicular), and placing the two parts of the upper portion disjoined by the perpendicular on the two sides of the lower portion (as in the annexed figure), an oblong is formed in which the half of the perpendicular is one side, and the base is the other. Wherefore half the perpendicular multiplied by the base is the area or number of equal compartments. Or, half the base multiplied by the perpendicular is just so much. In an obtuse-angled triangle also, the base multiplied by half the perpendicular is the area.—Gan.

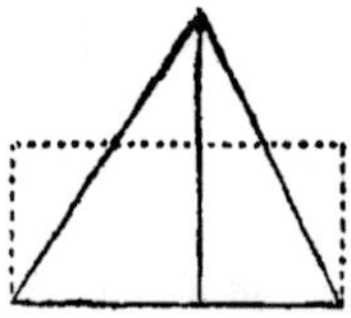

Statement: base 9; sides 10 and 17. By the rule in §163, the quotient found is 21. This cannot be subtracted from the base; wherefore the base is subtracted from it. Half the remainder is the segment, 6, and is negative, that is to say, in the contrary direction.[1] (See figure.) Thus the two segments are found 6 and 15. From which, both ways too, the perpendicular comes out 8. The area is 36.

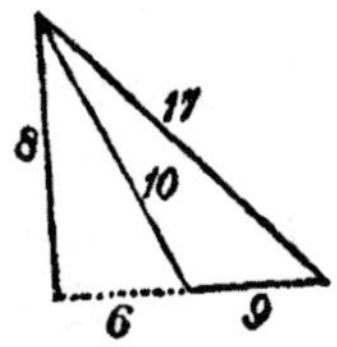

[As the commentators observe, the segments of the base made by the perpendicular in an obtuse-angled triangle are external, and their algebraical sum is their arithmetical difference.]

167. Rule.[2] Half the sum of all the sides is set down in four places; and the sides are severally

---

[The proof given by Súryadása is practically the same as the above, and so we omit it here.—Ed.]

[1] When the perpendicular falls without the base, as overpassing the angle in consequence of the side exceeding the base, the quotient found by the rule in § 163 cannot be taken from the base; for both origins of sides are situated in the same quarter from the fall of the perpendicular. Therefore subtracting the base from the quotient, half the residue is the segment and situated on the contrary side, being negative. Wherefore, as both segments stand on the same side, the smaller is comprehended in the greater, and, in respect of it, is negative. Thus all is cougruous an dunexceptionable—Gan. When the sum of the segments is to be taken, as they have contrary signs affirmative and negative, the difference of the quantities is that sum.—Súr. See *Vij-gan.* §5.

[2] For finding the gross area of a quadrilateral, and, by extension of the rule, the exact area of a triangle.—Gan. For finding the area by a method delivered by Srídhara—Rang.

subtracted. The remainders being multiplied together, the square root of the product is the area inexact in the quadrilateral, but pronounced exact in the triangle.[1]

[Let s denote the semi-perimeter of a triangle. Then area $=\sqrt{s(s-a)(s-b)(s-c)}$. See Todhunter's Trigonometry, Art. 247. This rule does not apply to the case of a quadrilateral, unless the quadrilateral is inscribable in a circle. The area of a quadrilateral in general can be shown to be $=\sqrt{\{(s-a)(s-b)(s-c)(sd-)-abcd\}}$, where a, b, c, d are the four sides, $2s$=perimeter and 2θ=sum of a pair of opposite angles. See Todhunter's Trigonometry revised by Hogg, Art. 256. In fact a quadrilateral in general is

---

[1] If the three remainders be added together, their sum is equal to half the sum of all the sides. The product of the continued multiplication of the three remainders being taken into the sum of those remainders, the product so obtained is equal to the product of the square of the perpendicular taken into the square of half the base. [It is not explained how this is the case. The last mentioned product =

$\frac{1}{4}a^2\left\{b^2-\frac{1}{4}\left(a+\frac{b^2-c^2}{a}\right)^2\right\}$, (taking a as base, and supposing $b>c$, and applying §163—164) $=\frac{4a^2b^2-(a^2+b^2-c^2)^2}{16}=\frac{1}{16}(a+b+c)(a+b-c)(a-b+c)$ $(b+c-a)=s\ (s-a)\ (s-b\ (s-c)=(s-a)\ (s-b)\ (s-c)\ (s-a+s-b+s-c)$ —ED.] It is a square quantity; for a square multiplied by a square gives a square. The square root being extracted, the product of the perpendicular by half the base is the result; and that is the area of the triangle. Therefore the true area is thus found. In a quadrilateral, the product of the multiplication does not give a square quantity, but an irrational one. Its approximate root is the area of the figure; not, however, the true one: for, when divided by the perpendicular, it should give half the sum of the base and summit—Súr. [The last remark does not hold good unless the quadrilateral be a trapezium.—ED.]

not determined by the four sides alone without an angle, whereas a triangle is determined by its three sides. Hence Bhaskara says that the expression derived from the above rule represents the gross area of a quadrilateral. See note to §§ 169—170. The proof of the rule given by Súryadása is not at all clear. See foot-note].

168. Example. In a quadrilateral figure, of which the base[1] is fourteen, the summit[2] nine, the flanks thirteen and twelve, and the perpendicular twelve, tell the area as it was taught by the ancients.

Statement: base 14; summit 9; sides 13 and 12; perpendicular 12. By the method directed, the result obtained is the surd 19800, of which the approximate root is somewhat less than 141. That, however, is not in this figure the true area. But, found by the method which will be set forth (§ 175), the true area is 138.

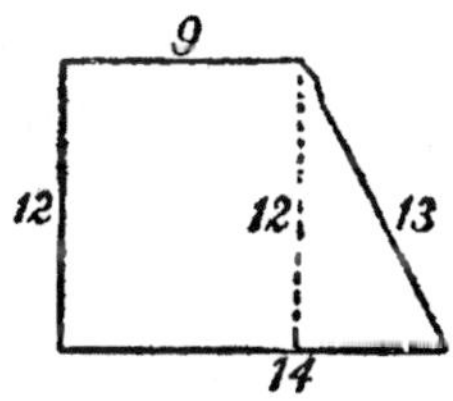

Statement of the triangle before instanced (§ 165).

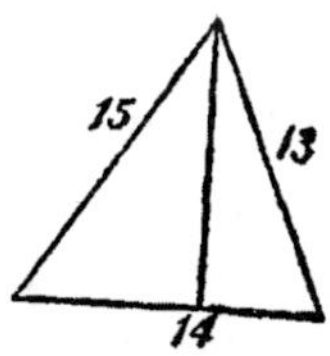

By the (present) method the area comes out the same, *viz.*, 84

169—170. Aphorism comprised in a stanza and a half. Since the diagonals of the quadrilateral are

---

[1] The greatest of the four sides is called the base—Gan. This definition is, however, too restricted. See §§178, 185.

[2] *Mukha*, *vadana*, or any other word denoting mouth; the side opposite to the base, the summit.

indeterminate, how should the area be in this case determinate ? The diagonals found as assumed by the ancients[1] do not answer in another case. With the same sides, there are other diagonals ; and the area of the figure is accordingly manifold.

For, in a quadrilateral, opposite angles being made to approach, contract their diagonal as they advance inwards: while the other angles receding outwards lengthen their diagonal. Therefore it is said that with the same sides there are other diagonals.

171. How can a person, neither specifying one of the perpendiculars, nor either of the diagonals, ask the rest ?[2] Or how can he demand a determinate area, while they are indefinite ?

172. Such a questioner is a blundering devil.[3] Still more so is he, who answers the question. For he considers not the indefinite nature of the lines[4] in a quadrilateral figure.

[The four sides alone without an angle do not determine the quadrilateral in general, and the area is consequently indeterminate. For, take AB as one side ; and with centre A and radius equal to another side describe a circle ; take *any point* D in the circumference, and join AD. With D, B as centres and radii equal to the

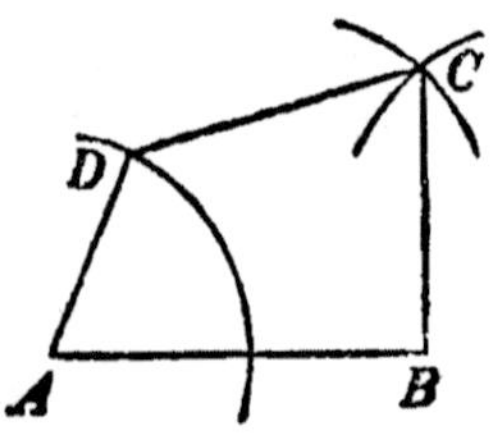

[1] By Srídhara and the rest.—Gan.
[2] The perpendiculars, diagonals, &c.—Gan.
[3] *Pisácha*, a demon or vampire ; so termed because he blunders.—Súr.
[4] Of the diagonal and perpendicular lines.—Súr.

other two sides, describe circles cutting each other in C. Join CB, CD. Then $ABCD$ is the quadrilateral which is indeterminate, since the angle BAD being not given, the point D may be taken anywhere in the circumference of the first described circle. Hence with the same sides, the diagonals may vary. But if the perpendicular from D to the line AB be given, or if the diagonal BD be given, it is easy to see that the point D becomes a fixed point in the circumference of the first described circle, and so the quadrilateral is determinate. In the case of a trapezium, it is easily seen that the four sides being given, the distance between two parallel sides is known, and so the figure is determinate. The rule in §167, however, is equally inapplicable to the case.]

173—175. Rule[1] in two and a half stanza. Let one diagonal of an equilateral tetragon be put as given. Then subtract its square from four times the square of the side. The square root of the remainder is the measure of the second diagonal.

The product of unequal diagonals multiplied together, being divided by two, will be the precise area in an equilateral tetragon. In a regular one with equal diagonals, as also in an oblong,[2] the product of the side and upright will be so.

In any other quadrilateral with equal prependiculars, the moiety of the sum of the base and summit, multiplied by the prependicular (is the area).

[1] In an equilateral tetragon, one diagonal being given, to find the second diagonal and the area; also in an equiperpendicular tetragon (trapezium) to find the area.—Gan. Equilateral tetragons are two-fold: with equal and with unequal diagonals. The first rule regards the equilateral tetragon with unequal diagonals (the rhombus).—Súr.

[2] *Ayata*, a long quadrilateral which has pairs of equal sides.—Gan.

[In an equilateral tetragon, the diagonals bisect each other at right angles. Hence (see figure), $x=\sqrt{b^2-a^2}$; $\therefore 2x=$ unknown diagonal $=\sqrt{4b^2-4a^2}$.

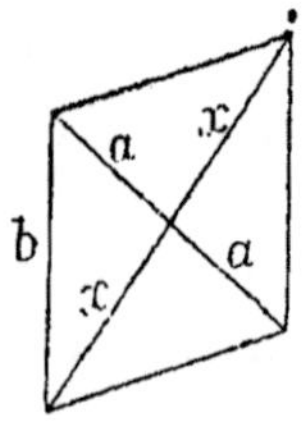

The area of the above figure is evidently equal to half the product of the diagonals. The other propositions stated above are well known elementary geometrical results. By "a quadrilateral with equal perpendiculars," the author means a trapezium.

176. Example. Mathematician, tell both diagonals and the area of an equilateral quadrangular figure whose side is the square of five: and the area of it, the diagonals being equal: also (the area) of an oblong, the breadth of which is six and length eight.

Statement of the first figure (rhombus). Here, assuming one diagonal 30, the other is found 40; and the area is 600.

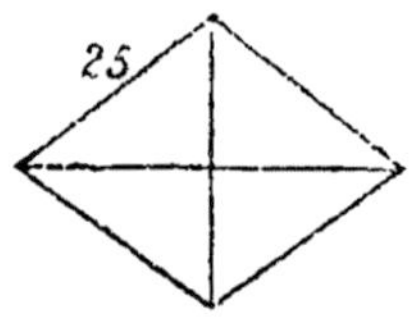

Or put one diagonal 14; the other is found 48; and the area is 336. See figure.

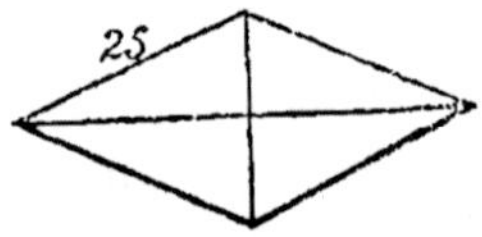

Statement of the second figure (square).

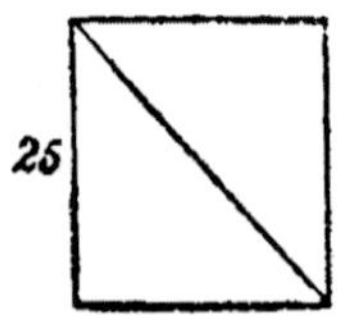

Here, taking the square root of the sum of the squares (§134), the diagonal comes out as the surd $\sqrt{1250}$, alike both ways. The area is 625.

Statement of the third figure (oblong). Area 48.

177. Example. Where the summit is eleven, the base twice as much as the summit, the flanks thirteen and twenty, and the perpendicular twelve; say what the area will be.

Statement:

The gross area (§ 167) is 250. The true area (§ 175) is 198.

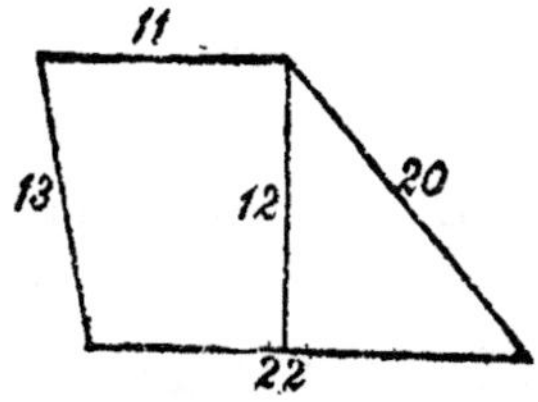

Or making three portions of the figure, and severally finding their areas, we get 30, 72, 96 (see figure); and summing up we get for the total area 198 as before.

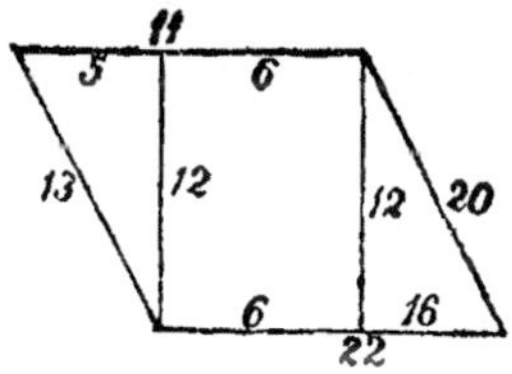

[The four sides of the trapezium being given, the perpendicular is necessarily known.]

178. Another example. Declare the diagonal, perpendicular and dimensions of the area, in a figure of which the summit is fifty-one, the base seventy-five, the left side sixty-eight, and the other side twice twenty.

179. Aphorism showing the connection of area, perpendicular and diagonal. If the perpendicular be known, the diagonal is so; if the diagonal be known, the perpendicular is so. If they be defi-

nite, the area is determinate. For, if the diagonal be indefinite, so is the perpendicular. Such is the meaning.

179 *continued.* Rule for finding the perpendicular.[1] In the triangle within the quadrilateral, the perpendicular is found as before taught (§§163—164); the diagonal and side being sides, and the base, a base.

Here, to find the perpendicular, a diagonal proceeding from the extremity of the left side to the origin of the right one is assumed to be 77; see figure. By this a triangle is constituted within the quadrilateral. In it that diagonal is one side, 77; the left side is another, 68; the base continues such, 75. Then, proceeding by the rule (§§163—164), the segments are found, $\frac{144}{5}$ and $\frac{231}{5}$; and the perpendicular, $\frac{308}{5}$.

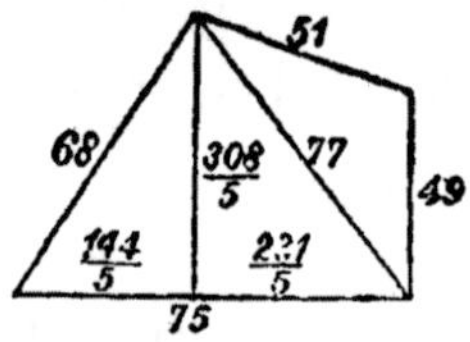

[The problem given is indeterminate, unless a diagonal, or an angle, or a perpendicular distance be given. So one diagonal is supposed to be 77. The process then adopted is the same as that shown in the note to §§ 163—164.]

180. Rule to find the diagonal, when the perpendicular is known.

The square root of the difference of the squares of the perpendicular and its adjoining side is pronounced the segment. The square of the base

[1] The diagonals being either given or assumed.—Gan.

less that segment being added to the square of the perpendicular, the square root of the sum is the diagonal.

In the above quadrilateral, the perpendicular from the extremity of the left side is put $\frac{308}{5}$. Hence the segment is found $\frac{144}{5}$; and by the rule (§180) the diagonal comes out 77.

[The reason for the rule is manifest.]

181—182. Rule to find the second diagonal: two stanzas.

In this figure, first a diagonal is assumed.[1] In the two triangles situated one on each side of the diagonal, this diagonal is made the base of each; and the other sides are given; the perpendiculars and segments[2] must be found. Then the square of the difference of two segments on the same side[3] being added to the square of the sum of the perpendiculars, the square root of the sum of those squares will be the second diagonal in all tetragons.[4]

---

[1] Either arbitrarily (see §183) or as given by the conditions of the question.—Gan.

[2] The two perpendiculars and the four segments.—Gan.

[3] Square of the interval of two segments measured from the same extremity.

[4] In the figure which is divided by the diagonal line, two triangles are contained, one on each side of that line; and their perpendiculars, which fall one on each side of the diagonal, are thence found. The difference between two segments on the same side will be the interval between the perpendiculars. It is taken as the upright of a triangle. Producing (see figure) one perpendicular by the addition of the other, (*i.e.*, drawing

In the same quadrilateral, the length of the diagonal passing from the extremity of the left side to the origin of the right one, is put 77. Within the figure cut by that diagonal line, two triangles are formed, one on each side of the diagonal. Taking the diagonal for the base of each, and the two other sides as given, the two perpendiculars and the several segments must be found by the method before taught. Thus the perpendiculars are found 24 and 60. Segments of the base made by the former, 45 and 32; those made by the latter, 32 and 45 Difference of the segments on the same side (that is, so much of the base as is intercepted between the perpendiculars) is 13. Its square 169. Sum of the perpendiculars 84. Its square 7056. Sum of the squares 7225. Square root of the sum 85. It is the length of the second diagonal. So in every like instance.

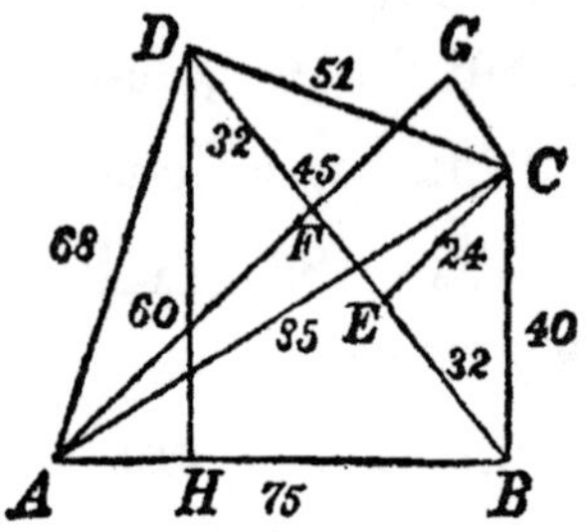

[The reason for the rule will be clear from the explanation given by Ganesa, cited in the foot-note. In

---

OG perpendicular to AF produced), the sum (AG) is made the side of the triangle. The second diagonal (AC) is hypotenuse. A triangle (AGC) is thus formed. From this is deduced, that the square root of the sum of the squares of the upright (which $= OG = EF = BF - BE$) and side (which $= AG = AF + CE$) will be the second diagonal: and the rule is demonstrated.—Gan.

In an equilateral tetragon, there is no interval between the perpendiculars; and the second diagonal is the sum of the perpendiculars.—*Ibid.*

the particular example chosen, it happens (see above figure) that $DF=EB$, but this need not always be the case. It also happens from the values of AB, BC, CA, that the angle ABC is a right angle]

183. Rule restricting the arbitrary assumption of a diagonal: a stanza and a half. The sum of the shortest pair of sides containing the diagonal being taken as a base, and the remaining two as the legs (of a triangle), the perpendicular is to be found: and, in like manner, with the other diagonal. The diagonal cannot by any means be longer than the corresponding base, nor shorter than the perpendicular answering to the other. Adverting to these limits, an intelligent person may assume a diagonal.

For a quadrilateral, contracting as the opposite angles approach, becomes a triangle; wherein the sum of the least pair of sides about one angle is the base, and the other two are taken as the legs. The perpendicular is found in the manner before taught. Hence the shrinking diagonal cannot by any means be less than the perpendicular; nor the other be greater than the base. It is so both ways. This, even though it were not mentioned, would be readily perceived by the intelligent student.

[What the author intends to say is (see figure to §§ 181-182) that the diagonal BD cannot be longer than $DC+BC$, but always shorter; nor can it be shorter than the perpendicular DH, but always longer. (Euclid, I.

20 and 19). The first sentence of the above section is meaningless; and so also is the proof given, *viz.*, "For a quadrilateral. contracting, &c."]

184. Rule to find the area: half a stanza. The sum of the areas of the two triangles on either side of the diagonal is assuredly[1] the area in this figure.

In the figure last specified, the areas of the two triangles are 924 and 2310. The sum of these is 3234, the area of the tetragon.

[The area of the triangle *BCD* (see figs. to §§ 181-182) is 924, and that of the triangle *ABD* is 2310.]

185—186. Rule: two stanzas. Making the difference between the base and summit of a (trapezium or) quadrilateral that has equal perpendiculars, the base (of a triangle), and the sides (its) legs, the segments of it and the length of the perpendicular are to be found as for a triangle. From the base of the trapezium subtract the segment, and add the square of the remainder to the square of the perpendicular; the square root of the sum will be the diagonal.

In a (trapezium or) quadrilateral that has equal perpendiculars, the sum of the base and least flank is greater than the aggregate of the summit and other flank.

[The rule gives the method of finding the diagonals of a trapezium the sides of which are given. It is demonstrated by Ganesa in the following manner :—

[1] It is the true and correct area, contrasted with the gross or inexact area of former writers.—Gan. and Súr.

Let the two triangles *ABE, DFC* be united into one triangle $A'B'C'$, their altitudes being equal. Then the altitude $A'E'$ of the new triangle $A'B'C'$ is the altitude of the trapezium, and the segments $B'E'$, $E'C'$ will be equal to *BE, FC*. Hence $AC^2 = AE^2 + EC^2 = A'E'^2 + (BC - B'E')^2$, which leads to the rule.

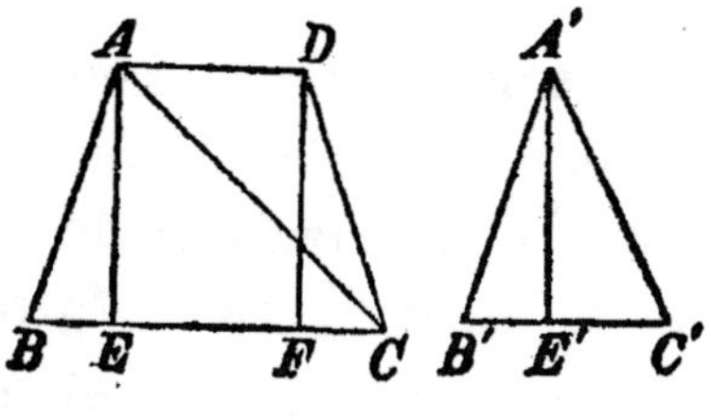

Also, $A'B' + B'C' > A'C'$,

$\therefore AB + BE + FC + EF > AD + DC$, ($\because AD = EF$)

i.e., $AB + BC \quad > AD + DC$.

AB need not be the 'least flank.']

187—189. Example. The sides measuring fifty-two and one less than forty, the summit equal to twenty-five, and the base sixty, this was given as an example by former writers for a figure having unequal perpendiculars; and definite measures of the diagonals were stated, fifty-six and sixty-three. Assign to it other diagonals, and those particularly which appertain to it as a figure with equal perpendiculars.

Statement.

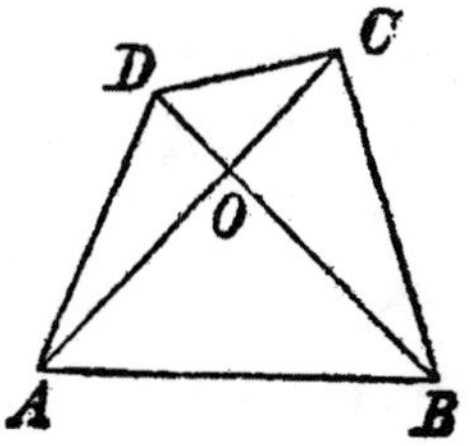

Fig. 1.

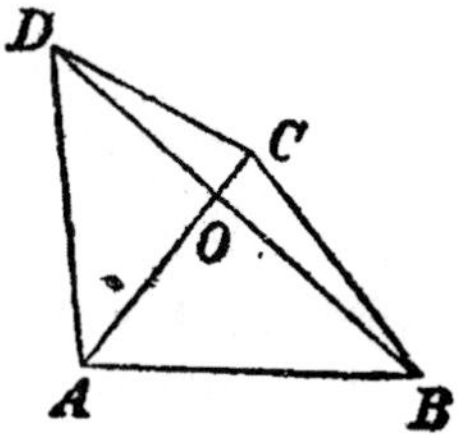

Fig. .

Here assuming one diagonal 63, the other is found as before, 56. Or, putting 32 instead of 56 for a diagonal (Fig. 2), the other, found by the process before shown, comes out in two portions, both surds, $\sqrt{621}$ and $\sqrt{2700}$. The sum of the roots (extracted by approximation) is the second diagonal $76\frac{22}{25}$.

[In Fig. 2, if we drop perpendiculars from B, D on AC, and find the segments of AC by § 163, we shall find that AC, BD intersect at right angles, and that $AO=30$.

Hence $BD=BO+DO=\sqrt{60^2-30^2}+\sqrt{39^2-30^2}$ $=\sqrt{2700}+\sqrt{621}=30\sqrt{3}+3\sqrt{69}=30\times1{\cdot}732...+3\times8{\cdot}307$ $...=51{\cdot}96...+24{\cdot}921...=76{\cdot}88...=76\frac{22}{25}$ nearly. The root is approximately extracted in the manner indicated in § 138.]

Again, if the above quadrilateral (Fig. 1) be one with equal perpendiculars, *i.e.*, a trapezium, (Fig. 3), consider the triangle $A'B'D'$ (Fig. 4), put to find the perpendicular DE, and the segments AE, FB, according to the rule in §§ 185-186. Here the segments are found $\frac{3}{5}$ and $\frac{172}{5}$; and the perpendicular, the surd $\sqrt{\frac{38016}{25}}$, of which the root found by approximation is $38\frac{622}{625}$. It is the equal perpendicular of the trapezium.

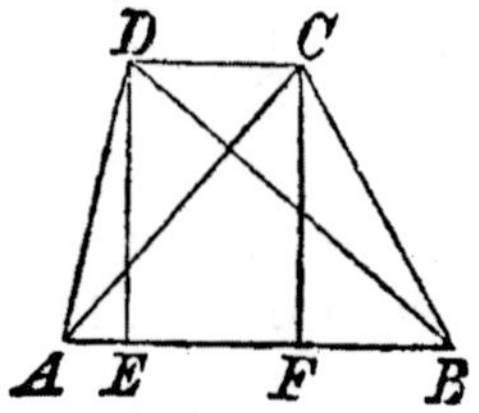

Fig. 3.

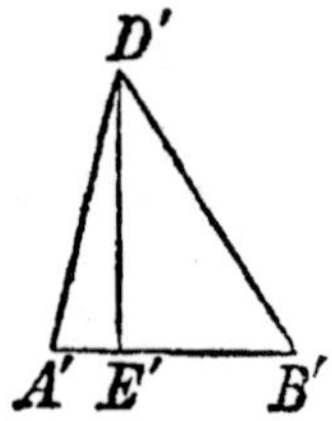

Fig. 4.

Next to find the sum of the squares of the perpendicular and difference between base and segment, we have, base of trapezium, 60 ; least segment $\frac{3}{5}$; difference $\frac{297}{5}$; square of the difference $\frac{88209}{25}$; square of the perpendicular $\frac{38016}{25}$; sum $\frac{126225}{25}$, or dividing by the denominator, 5049. It is the square of one diagonal (*BD*). Similarly, the square of the other diagonal *AC*) is 2176. Extracting the roots of these squares by approximation, the two diagonals come out $71\frac{1}{25}$ and $46\frac{16}{25}$.

In the above trapezium, the short side 39 added to the base 60 makes 99, which is greater than the aggregate of the summit and other flank, 77. Such is the limitation.

Thus, with the same sides, may be various diagonals in the tetragon. Yet, though indeterminate, diagonals have been sought as determinate, by Brahmagupta and other. Their rule is as follows :—

190. Rule.[1] The sums of the products of the sides about both the diagonals being divided by each other, multiply the quotients by the sum of the products of opposite sides ; the square roots of the results are the diagonals in a quadrilateral.

The objection to this mode of finding the diagonals is its operoseness, as I shall show by proposing a shorter method.

---

[1] A couplet cited from Brahmagupta, XII, 28.

[The rule applies only to a quadrilateral which can be inscribed in a circle. This, however, is not mentioned in the rule.

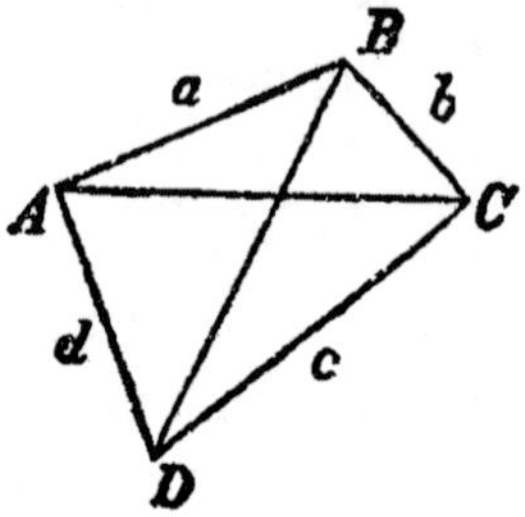

Let $ABCD$ be such a quadrilateral, and let $AB=a$, $BC=b$, $CD=c$, $DA=d$.

Then $AC^2=\dfrac{(ac+bd)(ad+bc)}{ab+cd}$

and $BD^2=\dfrac{(ac+bd)(ab+cd)}{ad+bc}$.

(See Todhunter's Trigonometry, Art. 254.) Thus the reason for the rule is obvious.

In the quadrilateral in §§ 187-189, Fig. 1, the sides are so taken that the diagonals intersect at right angles; and we easily find $OB=48$, $OD=15$ Thus $\cos(ODA)=\frac{15}{39}$ $=\frac{5}{13}$, and $\cos\ (OCB)=\frac{20}{52}=\frac{5}{13}$. Hence angle ODA = angle OCB, and therefore a circle passes round the quadrilateral. Consequently, the rule in § 190 will apply to this quadrilateral. It is probable that the rule was derived *a posteriori* from this particular instance, and not *a priori* from the fact of the quadrilateral being inscribable in a circle. The same quadrilateral is given as an example of the rule by Chaturveda Prithúdaka Swámí, in his commentary on Brahmagupta's treatise.]

191—192. Rule: two stanzas. The uprights and sides of two assumed right-angled triangles,[1] being multiplied by the reciprocal hypotenuses, become sides (of a quadrilateral): and in this

[1] Assumed conformably with the rule in § 145. An objection, to which Ganesa adverts and which he endeavours to obviate, is that this shorter method requires sagacity in the selection of assumed triangles; whereas the longer method is adapted to all capacities.

manner is constituted a quadrilateral, in which the diagonals are deducible from the two triangles.[1] The product of the uprights added to the product of the sides is one diagonal; the sum of the products of uprights and sides reciprocally multiplied, is the other.[2] When this short method exists, why an operose one was practised by former writers, we know not.

---

[1] This method of constructing a quadrilateral is taken from Brahmagupta XII, 38.

[2] A quadrilateral is divided into four triangles by its intersecting diagonals; and conversely, by the junction of four triangles, a quadrilateral is constituted. For that purpose, four triangles are assumed in this manner. Two triangles are first put in the mode directed (§ 145), the sides of which are all rational. Such sides, multiplied by any assumed number, will constitute other right-angled triangles, of which also the sides will be rational. By the twofold multiplication of hypotenuse, upright and side of one assumed triangle by the upright and side of the other, four (right-angled) triangles are formed, such that turning and adapting them and placing the multiples of the hypotenuses for sides, a quadrilateral is composed, (as shown below).

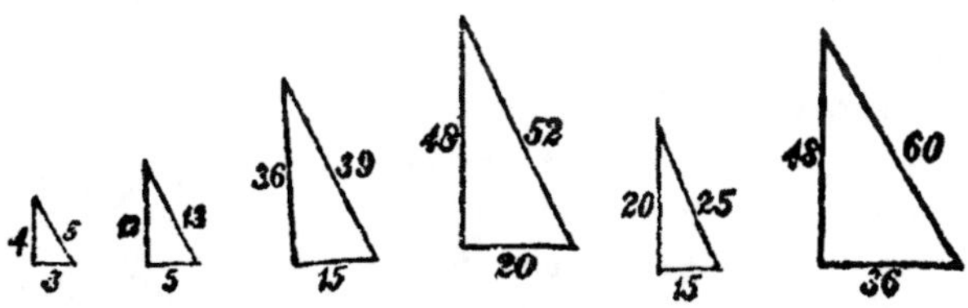

Here the uprights and sides of the arbitrary triangles (the first two on the left side), reciprocally multiplied by the hypotenuses, become sides of the quadrilateral; and hence the directions of the rule (§ 191). In a quadrilateral so constituted, it is apparent that the one diagonal (*AC*) is composed of two parts; one the product of

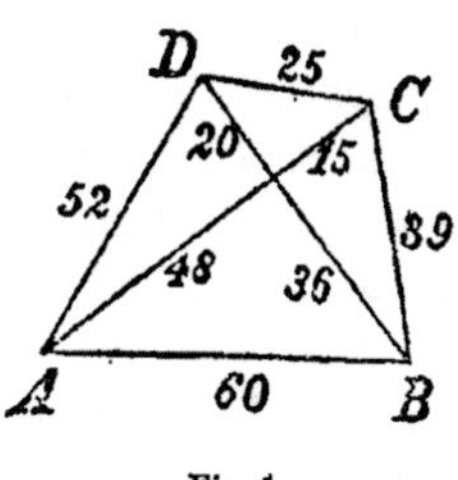

Fig. 1.

[This rule is of no importance whatever. It applies only to quadrilaterals constructed in the artificial manner indicated by the author, and fully explained by Ganesa; see foot-note. It is curious that the author, while censuring Brahmagupta's important rule (§190) as operose, entirely forgets that the rule propounded by himself is comparatively unimportant and of very limited application.]

Assuming two right-angled triangles, multiply the upright and side of one by the hypotenuse of the other: the greatest of the products is taken for the base; the least for the summit; and the other two for the flanks. (See Fig. 1 in the foot-note.)

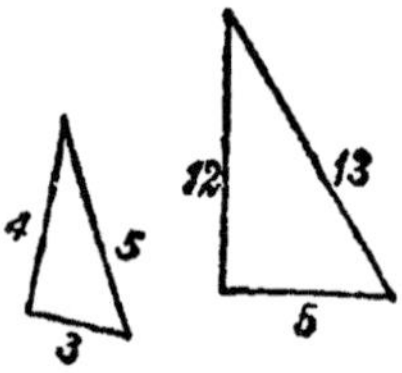

---

the uprights, the other the product of the sides of the arbitrary triangles. The other diagonal (*BD*) consists of two parts, *viz.*, the products of the reciprocal multiplication of uprights and sides. These two portions are the perpendiculars, for there is no interval between the points of intersection. This holds, provided the shortest side be the summit; the longest, the base; and the rest, the flanks. But if the component triangles be otherwise adapted, the summit and a flank change places, as in the adjoining figure. Here the two portions of the first diagonal as above found (*viz.*, 48 and 15) do not face, but are separated by an interval which is equal to the difference between the two portions (36 and 20) of the other diagonal, *viz.*, 16. It is the difference of two segments on the same side, found by a preceding rule (§§181-182), and is taken for the upright of a triangle as already explained (§§181-182, note); the sum of the two portions of the diagonal equal to the two perpendiculars is made the side. The square root of the sum of the squares of such upright and

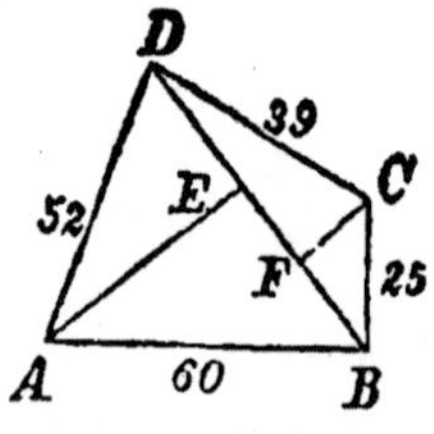

Here with much labour (by the former method) the diagonals are found 63 and 56.

With the same pair of right-angled triangles, the products of uprights and sides reciprocally multiplied are 36 and 20 ; the sum of which is one diagonal, 56. The products of uprights multiplied together, and sides taken into each other, are 48 and 15 ; their sum is the other diagonal, 63. Thus they are found with ease.

But if the summit and flank change places, and the figure be stated accordingly, the second diagonal will be the product of the hypotenuses of the two right-angled triangles, *viz.*, 65. (See Fig. 2 in the foot-note.)

---

side is equal to the product of the hypotenuses (13 and 5) : wherefore the author adds, "if the summit and flank change places, the *first* diagonal will be the product of the hypotenuses." (The MSS. have *first*, but Bháskara's text exhibits *second* instead.)

[This last remark is not clearly explained by the commentator. From the calculated values of AE and FB(see Fig. 2), we find that $\sin ABE=\frac{AE}{AB}=\frac{84}{60}=\frac{4}{5}$; and $\cos CBE=\frac{FB}{BC}=\frac{20}{25}=\frac{4}{5}$. Hence the angle ABC is a right angle.

And $\because AB=5\times 12$, and $BC=5\times 5$,

$\therefore AC=5\times 13$, and the truth of the remark is obvious. It need hardly be added that the remark applies only to quadrilaterals constructed in the artificial manner indicated by the author.—Ed.]

In like manner, for the tetragon before instanced (§178), to find the diagonals, a pair of rectangular triangles is put. Proceeding as directed, the diagonals come out 77 and 84. (See Fig. 3.)

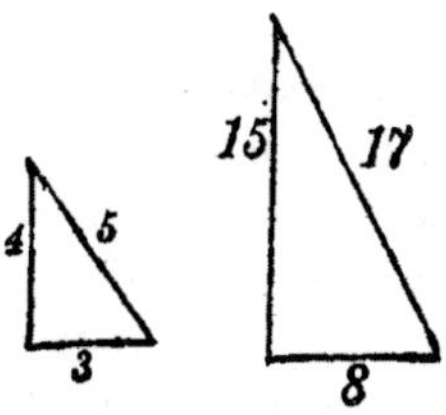

193—194. Example.[1] In a figure in which the base is three hundred, the summit a hundred and twenty-five, the flanks two hundred and sixty and one hundred and ninety-five, one diagonal two hundred and eighty and the other three hundred and fifteen, and the perpendiculars a hundred and eighty-nine and two hundred and twenty-four, what are the portions of the perpendiculars and diagonals below the intersections of them ? and the perpendicular let fall from the intersection of the diagonals, with the segments answering to it ? and the perpendicular of the needle formed by the prolongation of the flanks until they meet, as well as the segments corresponding to it ? and the

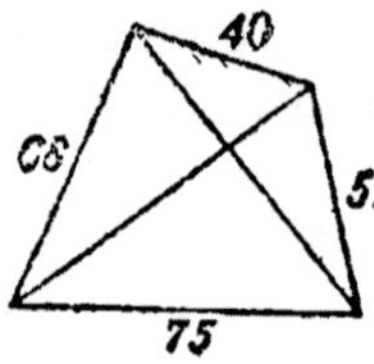

Fig. 3.

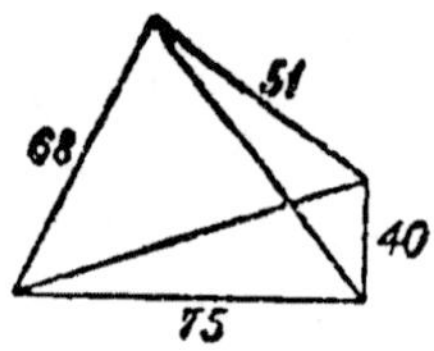

Fig. 4.

In the figure instanced, a transposition of the flank and summit takes place (see Fig. 4 which corresponds to the tetragon instanced in §178) ; wherefore the product of the hypotenuses (5 and 17) of the two right-angled triangles will be the second diagonal ; and they thus come out 77 and 85.—Gan.

[1] Having thus, from §173 to this place, shown the method of finding the area, &c., in the fourteen sorts of quadrilaterals, the author now exhibits another quadrilateral, proposing questions concerning segments produced by intersection.—Gan. For the instruction of the pupil, he exhibits the figure called (*súchí*) a needle.—*Mano.*

The problem is taken from Brahmagupta with a slight variation ; and this example differs from his only in the scale, his numbers being here increased five-fold. See Brahmagupta, XII, 32.

measure of both the needle's sides? All this declare, mathematician, if thou be thoroughly skilled in this (science of[1]) plane figure.

Statement. Length of the base 300. Summit 125. Flanks 260 and 195. Diagonals 280 and 315. Perpendiculars 189 and 224.

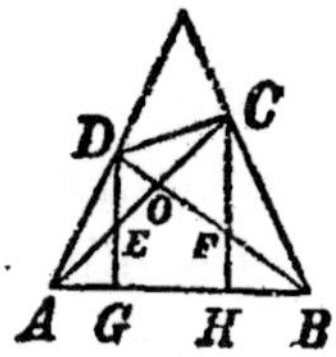

195—196. Rule: two stanzas. The interval between the perpendicular and its correspondent flank is termed the *sandhi*[2] or link of that perpendicular. The base less the link or segment is called the *pítha*[3] or complement of the same. The link or segment contiguous to that portion (of perpendicular or diagonal) which is sought, is twice set down. Multiplied by the other perpendicular in one instance, and by the diagonal in the other, and divided (in both instances) by the complement belonging to the other (perpendicular), the quotients will be the lower portions of the perpendicular and diagonal below the intersection.

Statement. Perpendicular 189. Flank contiguous to it 195. Segment intercepted between them (found by §134), 48. It is the link. The second segment is 252, and is called the complement.

[1] *Manoranjana.*

[2] *Sandhi,* union, alliance; connecting link.

[3] *Pítha,* lit. stool. Here the complement of the segment.

In like manner, the second perpendicular is 224. The flank contiguous to it, 260. Interval between them, being the segment called link, 132. Complement 168.

Now to find the lower portion of the first perpendicular 189. Its link 48 separately multiplied by the other perpendicular 224 and by the diagonal 280, and divided by the other complement 168, gives quotients 64, the lower portion of the perpendicular, and 80, the lower portion of the diagonal.

So for the second perpendicular 224, its link 132, severally multiplied by the other perpendicular 189 and by the diagonal 315, and divided by the other complement 252, gives 99 for the lower portion of the perpendicular, and 165 for that of the diagonal.

[The object of the rule is to find EG, EA, FH, FB, (See figure, §§ 193-194).

From similar triangles we have,

$$\frac{EG}{GA} = \frac{CH}{HA}, \text{ whence } EG = \frac{GA \times CH}{HA};$$

and $$\frac{AE}{AC} = \frac{GA}{HA}, \text{ whence } AE = \frac{GA \times AC}{HA}$$

Similarly for FH and FB.

Hence the reason for the rule is clear.]

197. Rule to find the perpendicular below the intersection of the diagonals.

The perpendiculars multiplied by the base and divided by the respective complements, are the erect poles : from which the perpendicular let fall from the intersection of the diagonals, as also the segments of the base, are to be found as before.[1]

Statement. Proceeding as directed, the erect poles are found 225 and 400. Whence, by a former rule (§159), the perpendicular below the intersection of the diagonals is deduced, 144 ; and the segments of the base 108 and 192.

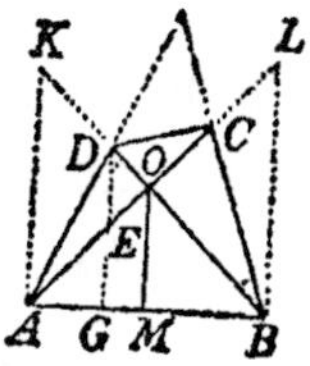

[The method employed is first to find AK, BL, the erect poles or perpendiculars on AB, and then to apply §159 to find OM. Now from similar triangles we have $\frac{AK}{AB}=\frac{GD}{GB}$, whence $AK=\frac{GD\times AB}{GB}$, with a similar value for BL. Hence the rule.]

198—200. Rule to find the perpendicular of the needle,[2] its legs and the segments of its base : three stanzas. The proper link multiplied by the other perpendicular and divided by its own, is termed the mean ;[3] and the sum of this and the opposite link is called the divisor.[4] Those two quantities, namely, the mean and the opposite link, being multiplied by the base and divided by that divisor, will be the respective segments of the

---

[1] By the rule in §159.

[2] *Súchí*, needle ; the triangle formed by the flanks of the quadrilateral until they meet.

[3] *Sama*, mean ; a fourth proportional to the two perpendiculars and the link or segment.

[4] *Hara*, divisor ; the sum of the *mean* and the other link or segment,

needle's base. The other perpendicular, multiplied by the base and divided by the divisor, will be the perpendicular of the needle. The flanks, multiplied by the perpendicular of the needle and divided by their respective perpendiculars, will be the legs of the needle. Thus may the subdivision of a plane figure be conducted by the intelligent, by means of the Rule of Three.

Here the perpendicular being 224, its link is 132. This multiplied by the other perpendicular, *viz.*, 189, and divided by its own, *viz.*, 224, gives the mean as it is named, $\frac{891}{8}$. The sum of this and the other link 48 is the divisor as it is called, $\frac{1275}{8}$. The mean and the other link severally taken into the base, being divided by this divisor, give the segments of the needle's base, $\frac{1536}{17}$ and $\frac{3564}{17}$. The other perpendicular 189, multiplied by the base and divided by the same divisor, yields the perpendicular of the needle, $\frac{6048}{17}$. The sides 195 and 260, multiplied by the needle's perpendicular and divided by their own perpendiculars respectively, *viz.*, 189 and 224, give the legs of the needle, which are the sides of the quadrilateral produced, *viz.*, $\frac{6240}{17}$ and $\frac{7020}{17}$.

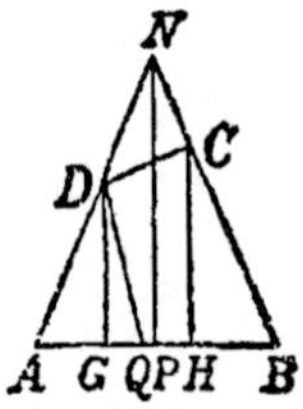

Thus in all instances under this head, taking the divisor for the argument, and making the multiplicand or multiplicator, as the case may be, the fruit or requisition, the Rule of Three is to be inferred by the intelligent mathematician.

[The object of the rule is to find PA, PB, NP, NA, NB (see above figure). It is demonstrated by Ganesa in the following manner :—Through D draw DQ parallel to NB, meeting AB, in Q. Then the triangles AQD and ABN are similar, and we evidently get $\frac{AB}{AQ}=\frac{AN}{AD}=\frac{NP}{DG}=\frac{PB}{GQ}=\frac{PA}{GA}$.........(1). Now $\frac{GQ}{DG}=\frac{HB}{CH}$ from similar triangles; thus $GQ=\frac{HB\times DG}{CH}$ = what the author calls *mean*, and consequently AQ or $mean+AG$ = what the author calls *divisor*. Thus finally, $PB=\frac{GQ\times AB}{AQ}$ from (1) $=\frac{mean\times \text{base}}{divisor}$; and $PA=\frac{GA\times AB}{AQ}$ from (1) $=\frac{\text{opposite link}\times\text{base}}{divisor}$. Again, from (1) $\frac{NP}{DG}=\frac{AB}{AQ}$, whence $NP=\frac{\text{other perpendicular}\times\text{base}}{divisor}$. Lastly, $\frac{NA}{DA}=\frac{NP}{DG}$, whence $NA=\frac{DA\times NP}{DG}$, and similarly for NB.]

201. Rule.[1] When the diameter of a circle[2] is multiplied by three thousand nine hundred and twenty-seven and divided by twelve hundred and

---

[1] To deduce the circumference of a circle from its diameter, and the diameter from the circumference.—Gan.

[2] *Vritta, vartula*, a circle. *Vyása, vishkambha, vistriti, vistára*, the breadth or diameter of a circle. *Paridhi, parináha, vritti nemi*, (and other synonyms of the felloe of a wheel), the circumference of a circle.

fifty, the quotient is the near[1] circumference : or multiplied by twenty-two and divided by seven, it is the gross circumference adapted to practice[2].

[Ganesa shows (see foot-note) that if the measure of the diameter of a circle be 1250, that of the side of a regular polygon of 384 sides inscribed in the circle will be very nearly 3927 (more accurately it will be $=\sqrt{98683}\times 12{\cdot}5 = 3926{\cdot}625\ldots$). This shows the degree of approximation of the fraction $\frac{3927}{1250}$ to the value of π. Converting the fraction into a decimal we get $3{\cdot}1416$, the true value of π being $3{\cdot}14159\ldots$]

202. Example. Where the measure of the diameter is seven, friend, tell the measure of the circumference : and where the circumference is twenty-two, find the diameter.

Statement : diameter 7.

Answer : circumference $21\frac{1239}{1250}$, or gross circumference 22.

Statement : circumference 22.

Reversing multiplier and divisor, the diameter comes out $7\frac{11}{3927}$; or gross diameter 7.

---

[1] *Sûkshma*, delicate or fine, nearly precise; contrasted with *sthûla*, gross, or somewhat less exact, but sufficient for common purposes.—Gang. and Súr.

Brahmagupta puts the ratio of the circumference to the diameter as three to one for the gross value, and takes the square root of ten times the square of the diameter for the neat value of the circumference. See Brahma. XII, 40. [This is more rough even than $\frac{22}{7}$; for $\sqrt{10}=3{\cdot}1622\ldots$, and $\frac{22}{7}=3.14285\dot{7}$.—Ed.]

[2] As the diameter increases or diminishes, so does the circumference increase or diminish : therefore to find the one from the other, make proportion, as the diameter of a known circle is to the known circum-

203. Rule. In a circle, a quarter of the diameter multiplied by the circumference is the area. That multiplied by four[1] is the net all around the ball.[2]

---

ference, so is the given diameter to the circumference sought; and conversely, as the circumference is to the diameter, so is the given circumference to the diameter sought.

Further, the semi-diameter is equal to the side of a regular hexagon within the circle, as will be shown. From this the side of a regular dodecagon may be found in this manner :— the semi-diameter being hypotenuse, and half the side of the hexagon, the side, the square root of the difference of their squares is the upright : subtracting which from the semi-diameter, the remainder is the arrow (or height of the arc). Again, this arrow being the upright and half the side of the hexagon, a side, the square root of the sum of their squares is the side of the dodecagon.

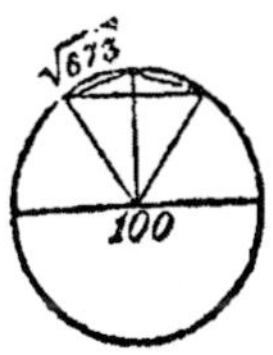

From this, in like manner, may be found the side of a polygon of 24 sides : and so on, doubling the number of sides in the polygon, until the side be near to the arc. The sum of such sides will be the circumference of the circle nearly, Thus, the diameter being 100, the side of the dodecagon is the surd $\sqrt{673}$; and that of a polygon of 384 sides is nearly equal to the arc. By computation it comes out the surd $\sqrt{98683}$. Now the proportion, if to the square of the diameter put at 100, *viz.*, 10000, this be the square of the circumference, *viz.*, 98683, then to the square of the assumed diameter 1250, *viz.*, 1562500, what will be the circumference ? Answer : the square root 3927 without remainder.—Gan.

[1] Or rejecting equal multiplier and divisor, the circumference multiplied by the diameter is the surface.—Gan.

[2] *Prishtha-phala,* superficial content : compared to the net formed by the string with which cloth is tied to make a playing ball. *Ghana-phala,* solid content : compared to a cube, and denominated from it, cubic.

This content of the surface of the sphere, multiplied by the diameter and divided by six, is the precise solid, termed cubic, content within the sphere.[1]

[This rule gives the well-known expressions for the area of a circle, and the surface and volume of a sphere. Ganesa gives interesting but rough demonstrations in the case of the area of a circle and the volume of a

---

[1] Dividing the circle into two equal parts, cut the content of each into any number of equal angular spaces, and expand it so that the circumference may

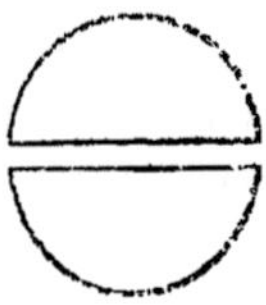

become a straight line as in the adjoining figure. Then let the two portions approach so that the angular spaces of the one may enter into the similar intermediate vacant spaces of the other, as in the figure, thus constituting an oblong, of which the semi-diameter is one side, and half the circumference the other. The product of their multiplication is the area. Half by half is a quarter. Therefore a quarter of the diameter multiplied by the circumference is the area.—Gan.

See in the *Goládhyáya* of the *Siddhánta-siromani*, a demonstration of the rule, that the surface of a sphere is four times the area of a great circle, or equal to the circumference multiplied by the diameter.—*Ibid.* [See the *Goládhyáya*, Wilkinson's translation, III, 52.—Ed.]

To demonstrate the rule for the solid content of a sphere, suppose the sphere divided into as many little pyramids, or long needles with an acute tip and square base, as in the number by which the surface is measured, the height of each pyramid being equal to the radius of the sphere. The base of each pyramid is a unit of the scale by which the dimensions of the surface are reckoned; and the altitude being a semi-diameter, one-third of their product is the content: for a needle-shaped excavation is one-third of an excavation in the form of a rectangular parallelopiped of the same base and height, as will be shown (§221). Therefore (unit taken into) a sixth part of the diameter is the content of one such pyramidical portion: and that multiplied by the surface gives the solid content of the sphere.—*Ibid.*

sphere, and refers to the *Goládhyáya* for the case of the surface of a sphere. See foot-note.]

204. Example. Intelligent friend, if thou know well the spotless *Lílávatí*, say what the area of a circle is, the diameter of which is measured by seven : and the surface of a globe, or area like a net upon a ball, the diameter being seven : and the solid content within the same sphere.

Statement : diameter 7.

Answer : area of the circle, $38\frac{2423}{5000}$. Superficial content of the sphere, $153\frac{1173}{1250}$. Solid content of the sphere, $179\frac{1487}{2500}$.

205—206. Rule : a stanza and a half. The square of the diameter being multiplied by three thousand nine hundred and twenty-seven, and divided by five thousand, the quotient is the nearly precise area ; or multiplied by eleven and divided by fourteen, it is the gross area adapted to common practice. Half the cube of the diameter, with its twenty-first part added to it, is the solid content of the sphere.

The area of the circle, nearly precise, comes out as before $38\frac{2423}{5000}$, or gross area $38\frac{1}{2}$. Gross solid content $179\frac{2}{3}$.

[The area of a circle $= \frac{1}{4}\pi d^2$, d being the diameter

$$= \frac{3927}{4 \times 1250}d^2 = \frac{3927}{5000}d^2 \; ;$$

or more roughly $= \frac{22}{4 \times 7}d^2 \quad = \frac{11}{14}d^2$.

Again, the volume of a sphere $=\frac{4}{3}\pi\frac{d^3}{8}$, d being the diameter

$$=\frac{22}{7\times6}d^3=\frac{11}{21}d^3 \text{ roughly}$$

$$=\frac{1}{2}d^3\left(1+\frac{1}{21}\right).$$

Thus the reasons for the rules are obvious.]

206—207. Rule[1] : a stanza and a half. The sum and difference of the chord and diameter being multiplied together, and the square root of the product being subtracted from the diameter, half the remainder is the arrow.[2] The diameter less the arrow being multiplied by the arrow, twice the square root of the product is the chord. The square of half the chord being divided by the arrow, the quotient added to the arrow is pronounced to be the diameter of the circle.

[By the word *arrow* is meant the height of the arc.

Súryadása gives the following proof of the formula for the arrow.

Let AB be the chord, CD, the arrow and O the centre. Join BO, and produce it to meet the circumference in E. Draw the chord EGF perpendicular to the diameter DH. Join BF.

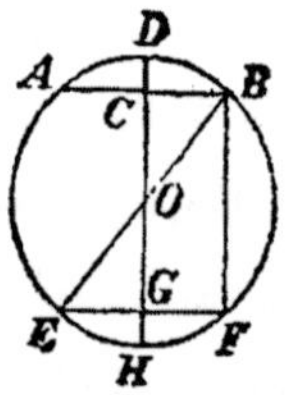

[1] In a circle cut by a right line, to find the chord, arrow, &c.; that is, either the chord, the arrow, or the diameter being unknown, and the other two given, to find the one from the others.—Gan, and Súr.

[2] A portion of the circumference is a bow (*dhanush, chápa*). The right line between its extremities, like the string of a bow is its chord (*jívá, jyá, guna; maurví*). The line between them is the arrow (*sara, ishu*), as resembling one set on a bow.—Gan. and Súr.

Then $CD=\frac{1}{2}(DH-CG)=\frac{1}{2}(DH-BF)$

$$=\frac{1}{2}(DH-\sqrt{BE^2-EF^2})=\frac{1}{2}(DH-\sqrt{DH^2-AB^2}),$$

whence the rule.

Again, $CH.\ CD=AC^2$ (Euclid III. 35)

$$\therefore AB=2AC=2\sqrt{CH.\ CD}.$$

Also $\frac{AC^2}{CD}=CH$;

$$\therefore DH=\frac{AC^2}{CD}+CD.]$$

208. Example. In a circle of which the diameter is ten, the chord being measured by six, say friend what the arrow is : and from the arrow tell the chord ; and from chord and arrow, the diameter.

Statement : diameter 10. When the chord is 6, the length of the arrow comes out 1.

Or, the arrow being 1, the chord is found 6. Or from the chord and arrow the diameter is deduced 10.

209—211. Rule :[1] three stanzas. By 103923, 84853, 70534, 60000, 52055, 45922, and 41031, multiply the diameter of a circle, and divide the respective products by 120000 ; the quotients are severally, in their order, the sides of polygons

[1] To find the sides of regular inscribed polygons.—Gan. and Súr.

from the triangle to the enneagon (inscribed) within the circle.[1]

[In this rule the author gives the fractions by which the diameter of a circle is to be severally multiplied, in order to get the sides of inscribed regular figures from the triangle to the enneagon. Ganesa shows by a purely geometrical method how the fractions are arrived at, in the case of the triangle, the square, the hexagon, and the octagon; and remarks that a similar proof cannot be given in the case of the pentagon, the heptagon and the enneagon. See foot-note. Súryadása tries to supply the proofs in these cases, but his attempt is a failure; for the proofs he gives are not at all rigid and satisfactory, and it is not worth while to give them in the foot-note, as Colebrooke does. By the help of trigonometrical tables, proofs in all the cases may be given in a general manner as follows :—

Let a denote the side of a regular polygon of n sides inscribed in a circle of radius r. Then $a=2r \sin \frac{\pi}{n}$ (see

---

[1] Describe a circle with any radius at pleasure, divide the circumference into three equal parts and mark the points; and with those points (A, B, C) as centres and with the same radius, describe three circles, which will be equal in circumference to the first circle; and it is thus manifest that the side of the regular hexagon within the circle is half a diameter.

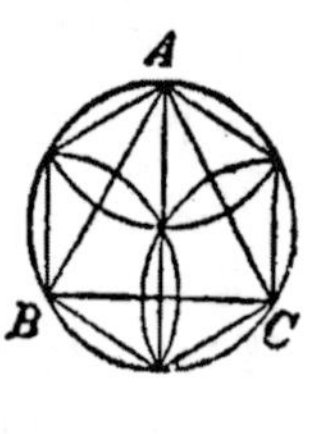

The side of an equilateral triangle (inscribed) in a circle is the upright, the diameter is hypotenuse and the side of the hexagon is side of a right-angled triangle. See above figure. Therefore the square root of the difference of the squares of the semi-diameter and diameter is the side of the (inscribed) equilateral triangle: *viz.*, for the proposed diameter (120000), 103923.

Todhunter's Trigonometry, Art. 255). Hence, side of inscribed equilateral triangle $=2r \sin 60° = 2r \frac{\sqrt{3}}{2} = 2r \frac{1 \cdot 7320508...}{2} = 2r \times \cdot 8660254...$ Now the fraction $\frac{103923}{120000}$ $= \cdot 866025$; thus the approximation is very close.

The side of the inscribed square $=2r \sin \frac{\pi}{4} = 2r \frac{\sqrt{2}}{2} = 2r \times$ $\cdot 7071067...$

Now the fraction $\frac{84853}{120000} = \cdot 708108\dot{3}$; thus the fraction is a little too large.

The side of the inscribed pentagon $=2r \sin 36° = 2r$ $\times \cdot 5877853$ from a table of natural sines. Now the fraction $\frac{70534}{120000} = \cdot 58778\dot{3}$; thus the fraction is a little too small.

The side of the inscribed hexagon is equal to the radius.

The side of the inscribed heptagon $=2r \sin (25° \; 42' \; 51'')$ nearly $=2r \times \cdot 4338819$ from the tables and the theory of proportional parts. Now the fraction $\frac{52055}{120000} =$ $\cdot 433791\dot{6}$; thus the fraction is a little too small.

The side of a square is hypotenuse, and the semi-diameter is upright and side. Wherefore the square root of twice the square of the semi-diameter is the side of the (inscribed) square: *viz.*, for the diameter assumed, 84853.

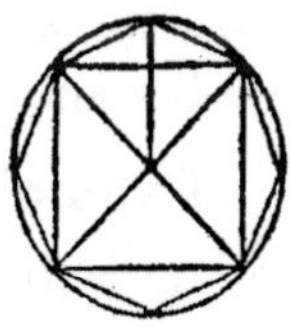

The side of the regular octagon (see above figure) is hypotenuse, half the side of the square is upright, and the difference between that and the semi-diameter is the side. Wherefore the square root of the sum of the squares of half the side of the square and the semi-diameter less half the side of the square is the side of the (inscribed) regular octagon: *viz.*, for the diameter as put, 45922.

The proof of the sides of the regular pentagon, heptagon and enneagon cannot be shown in a similar manner.—Gan.

The side of the inscribed octagon $= 2r \sin 22\frac{1}{2}^\circ = 2r \times \cdot 3826834$ from the tables. Now the fraction $\frac{45922}{120000} = \cdot 38268\dot{3}$; thus the approximation is very close.

The side of the inscribed nonagon $= 2r \sin 20^\circ = 2r \times \cdot 3420201$ from the tables. Now the fraction $\frac{41031}{120000} = \cdot 341925$; thus the fraction is a little too small.

In the appendix to the *Goládhyáya*, called *Jyotpatti*, Bháskara has given an elaborate method of constructing the sines of various angles, adopting the old definition of the sine. (See Trigonometry, Art. 71.) The values deduced by his method closely approximate the values given in our modern tables, there being slight discrepancies in some cases, which account for the discrepancies noticed above between the values of the sides of some of the inscribed regular polygons as given in the text, and their values as calculated from the tables. A table of sines and versed sines of certain angles in arithmetical progression is also given in the *Súrya-siddhánta*. I. 15—27, the values there stated being less accurate than those deducible from Bháskara's method. See *Súrya-siddhánta*, Bápu Deva Sástrí's translation, II. 16, foot-note. The decimal notation is nowhere used either by Bháskara or in the *Súrya-siddhánta*. See note to § 138.

212. Example. Within a circle of which the diameter is two thousand, tell me severally the sides of the inscribed equilateral triangle and other polygons.

Statement : diameter 2000.

Answer : side of the triangle, $1732\frac{1}{20}$; of the tetragon, $1414\frac{13}{60}$; of the pentagon, $1175\frac{17}{30}$; of the hexagon ; 1000 ; of the heptagon, $867\frac{7}{12}$; of the octagon, $765\frac{11}{30}$; of the nonagon, $683\frac{17}{20}$.

From variously assumed diameters, other chords are deducible, as will be shown by us under the head of construction of sines (*Jyotpatti*) in the treatise on Spherics.

[See *Goládhyáya*, appendix, Wilkinson's translation.]

The following rule teaches a short methed of finding the gross chords

213. Rule. The circumference less the arc being multiplied by the arc, the product is termed first.[1] From the quarter of the circumference multiplied by five, subtract that first product, and by the remainder divide the first product taken into four times the diameter. The quotient will be the chord.

[This rule, as the author himself observes, gives a method of finding approximately the chords of given arcs of a circle. The commentators give an unsatisfactory and almost fanciful demonstration of the rule The nature of the approximation may be shown thus :—

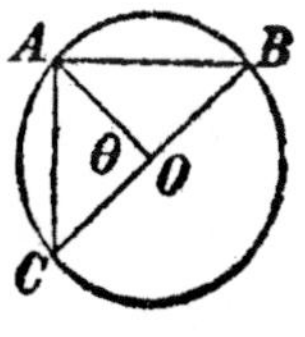

Let AB be the given arc whose chord is to be found. Draw the diameter BOC, and join AC, AO. Let θ denote the angle AOC, r the radius, and c the circumference of the circle. Then the value of the chord AB as given by the rule

$$=\frac{\text{arc } ABC \times \text{arc } AB \times 8r}{\frac{5}{4}c^2 - \text{arc } ACB \times \text{arc } AB}$$

$$=\frac{8r\left\{(\frac{1}{2}c)^2 - (\text{arc } CA)^2\right\}}{\frac{5}{4}c^2 - (\frac{1}{2}c)^2 + (\text{arc } CA)^2} = \frac{2r(4\pi^2 - 4\theta^2)}{4\pi^1 + \theta^2}$$

---

[1] *Prathama, ádya*, first (product).

$$=2r.\frac{1-\left(\frac{\theta}{\pi}\right)^2}{1+\left(\frac{\theta}{2\pi}\right)^2}=2r.\left\{1-\left(\frac{\theta}{\pi}\right)^2\right\}\left\{1-\left(\frac{\theta}{2\pi}\right)^2+\&c.\right\}$$

$=2r.\left(1-\frac{5\theta^2}{4\pi^2}\right)$, neglecting powers of $\frac{\theta}{\pi}$ beyond the second. The accurate value of the chord $AB=2r\cos\frac{\theta}{2}=$ $2\pi\left(1-\frac{\theta^2}{8}+\&c.\right.$. Taking $\pi=\frac{22}{7}$, the value of the fraction $\frac{5}{4\pi^2}$ will be found to be very slightly greater than $\frac{1}{8}$. This shows the nature of the approximation. It is worthy of notice that the rule gives the exact value of the chord when the arc is a sixth part of the circumference. The rule appears to have been obtained empirically after repeated trials.]

214. Example. Where the semi-diameter is a hundred and twenty, and the arc of the circle is measured by an eighteenth multiplied by one and so forth (up to nine[1]), tell quickly the chords of those arcs.

Statement : diameter 240.

Here the circumference is 754 (nearly).

Arcs being taken which are multiples of an eighteenth of the circumference, the (corresponding) chords are to be sought.

Or for the sake of facility, abridging both circumference and arcs by the eighteenth part of the

[1] Up to nine, or half the number of arcs; for the chords of the eighth and tenth will be the same, and so will those of the seventh and eleventh, and so forth.—Gan.

circumference, the same chords are fo und. Thus, circumference, 18 ; arcs, 1, 2, 3, 4, 5, 6, 7, 8, 9. Proceeding as directed, the chords come out 42, 82, 120, 154, 184, 208, 226, 236, 240.

In like manner, with other diameters (chords of assigned arcs may be found).[1]

[We can easily test the accuracy of the values of the chords given above by calculating their actual values from a table of natural sines. It will be found that the values given are in some cases less, and in others greater than the true values.]

215. Rule.[2] The square of the circumference is multiplied by a quarter of the chord and by five, and divided by the chord added to four times the diameter ; the quotient being subtracted from a quarter of the square of the circumference, the square root of the remainder, taken from half the circumference, will leave the arc.

[This rule is derived from the preceding one. Denoting the arc AB (see figure in the note to §213) by x, we get $AB = \frac{8r\,(c-x)\,x}{\frac{5}{4}c^2 - (c-x)x}$ (§213), whence $x^2 - cx + \frac{5\,AB.\,c^2}{4\,(8r+AB)} = 0$;

$\therefore\ \frac{c}{2} - x = \sqrt{\left\{\frac{c^2}{4} - \frac{5\,AB\,c^2}{4(8r+AB)}\right\}}$, the upper sign being taken, as x is supposed to be less than the semi-circumference. Hence the reason for the rule is obvious.

The following empirical and approximate rule for finding the arc is cited by Ganesa from Áryabhatta :— Six times the square of the arrow being added to the

[1] Gang. &c,

[2] To find the arc from the chord given.

square of the chord, the square root of the sum is the arc. If 2θ denote the angle subtended by the arc at the centre, and r the radius, the expression for the arc as given by the rule

$$= \sqrt{\{4r^2 \sin^2\theta + 6r^2(1-\cos\theta)^2\}}$$
$$= r\sqrt{\{10 + 2\cos^2\theta - 12\cos\theta\}}$$
$$= r\sqrt{\left\{10 + 2\left(1 - \frac{\theta^2}{2} + \ldots\ldots\right)^2 - 12\left(1 - \frac{\theta^2}{2} + \ldots\ldots\right)\right\}}$$

$= r.\,2\theta$, (neglecting higher powers of θ), which is the true value of the arc. This shows the nature of the approximation involved in the rule. In the case of the semi-circle, the rule gives for the length of the arc the expression $\sqrt{10}\,r$, so that the value of π is assumed to be $\sqrt{10}$.]

216. Example. From the chords which have been here found, now tell the length of the arcs, if, mathematician, thou have skill in computing the relation of arc and chord.

Statement: chords, 42, 82, 120, 154, 184, 208, 226, 236, 240.

Circumference abridged 18. The arcs thence found are, 1, 2, 3, 4, 5, 6, 7, 8, 9. They must be multiplied by the eighteenth part of the circumference.[1]

[1] Súryadása and Gangádhara notice other figures omitted by the author, *e.g.*, *gaja-danta* or elephant's tusk, which may be treated as a triangle according to Srídhara ; *bálendu* or crescent, which may be considered as composed of two triangles, according to the same author ; *yava* or barleycorn, a convex lens, treated as consisting either of two triangles or two bows, according to Gangádhara ; *nemi* or felloe ; *vajra* or thunderbolt, treated as a quadrilateral with two bows, according to Gangádhara ; *sankha* or conch ; *mridanga* or great drum ; and several others.

[Bháskara has given no rule for finding the area of a bow or segment of a circle. Ganesa cites in his commentary two rules for this purpose, which are practically the same. One of them, due to his father Kesava, is as follows :—The arrow being multiplied by half the sum of the chord and arrow, and a twentieth part of the product being added, the sum is the area of the segment. The other rule due to Srídhara is as follows :—The square of the arrow multiplied by the square of half the sum of the chord and arrow, being multiplied by ten and divided by nine, the square root of the product is the area of the bow. Since the fraction $\frac{21}{20}$ is very nearly equal to the fraction $\frac{\sqrt{10}}{3}$, we see that these two rules are practically the same. They both appear to be empirical and give very rough results, as may be readily seen by applying them to one or two particular cases. Thus, taking the first rule and applying it to the case when the segment is a semi-circle, we get for the area, the expression $\frac{21}{20}\frac{3r^2}{2}$, the true area being $\frac{\pi r^2}{2}$, r denoting the radius; so that in this case the value of π is taken to be $\frac{63}{20}$, which is greater than the true value. Again, applying the same rule to the case when the arc of the segment is a quadrant, we get for the area, the expression $\frac{21}{20}.\frac{r^2}{4}$, the true area being $\frac{(\pi-2)r^2}{4}$; so that in this case the value of π is taken to be $\frac{61}{20}$, which is less than the true value. Ganesa himself gives the accurate method of finding the area of the segment, namely, by subtracting the area of the triangle formed by the radii and the chord of the segment from the area of the sector. The same rule is also given in the *Manoranjana.*]

CHAPTER VII.

EXCAVATIONS[1] AND CONTENTS OF SOLIDS.

217—218. Rule[2] : a couplet and a half. Taking the breadth in several places,[3] let the sum of the measures be divided by the number of places : the quotient is the mean measure. So likewise with the length and depth.[4] The area of the plane figure, multiplied by the depth, will be the number of solid cubits contained in the excavation.

[The rule is very rough, giving a result much smaller than the true one. It is curious that such a rough rule was given when the author intended to lay down the correct rule immediately afterwards (§221). The tank contemplated is no doubt an ordinary one with slant sides, and we need not take the measurements in several places ; the length and breadth of the mouth and bottom, and the depth of the bottom from the mouth, being sufficient for finding the volume accurately. See §221.]

---

[1] *Kháta-vyavahára*. The author treats first of excavations, secondly of stacks of bricks and the like, thirdly of sawing of timber, and fourthly of stores of grain, in as many distinct chapters.

[2] For measuring an excavation, the sides of which are trapezia.—Gan.

[3] *Vistára*, breadth ; *dairghya*, length ; *bedha*, depth. *Kháta*, an excavation ; *sama-kháta*, a cavity in the form of a rectangular parallelopiped, cylinder, &c. ; *vishama-kháta*, a cavity in the form of an irregular solid; *súchi-khata*, an acute one, a pyramid or cone. *Sama-mili*, mean measure. *Ghana-phala*, the content of the excavation.

[4] The irregular solid is reduced to a regular one, to find its content. —Súr.

219—220. Example : two stanzas. Where the length of the cavity, owing to the slant of the sides, is measured by ten, eleven and twelve cubits in three several places, its breadth by six, five and seven, and its depth by two, four and three : tell me, friend, how many solid cubits are contained in that excavation.

Statement :	lengths,	12	11	10 ;
	breadths,	7	5	6 ;
	depths,	3	4	2.

Here finding the mean measure, the breadth is 6 cubits, the length 11, and the depth 3. The number of solid cubits is found, 198.

221. Rule[1]: a couplet and a half. The aggregate of the areas at the top and at the bottom, and of that resulting from the sum (of the sides of the summit and base), being divided by six, the quotient is the mean area : that multiplied by the depth is the neat[2] content.[3] A third part of the content

[1] To find the content of a prism, pyramid, cylinder and cone.

[2] Contrasted with the result of the preceding rule, which gave a gross or approximate measure.

[3] Half the sum of the breadths at the mouth and bottom is the mean breadth ; and half the sum of the lengths at the mouth and bottom is the mean length : their product is the area at the middle of the tank. (Four times that is the product of the sums of the length and breadth.) This, added to once the area at the mouth and once the area at the bottom, is six times the mean area.—Gan.

of the regular equal solid is the content of the acute one.[1]

[This rule gives the exact volume. The tank contemplated is an ordinary one with uniformly slanting sides. Let $ABCD$ be the mouth of the tank, and $EFGH$ its base, both being supposed rectangular. Suppose the mouth of the tank to be covered by a plane. Draw perpendiculars on this plane from E, F, G, H, and from the feet of these perpendiculars, draw perpendiculars on AB, BC, CD, DA, as in the figure. Let $AD=a$, $AB=b$, $EH=c$, $EF=d$, and z=vertical depth of the tank. Then it will be easily seen that the tank is divided into a rectangular parallelopiped whos volume is cdz; four triangu lar prisms, two and two being equal, the united volume being $\frac{1}{2}z\{(a-c)d+(b-d)c\}$; and four equal pyramids on square bases, one at each corner, the united volume being $\frac{1}{3}z(a-c)(b-d)$. Hence the volume of the tank

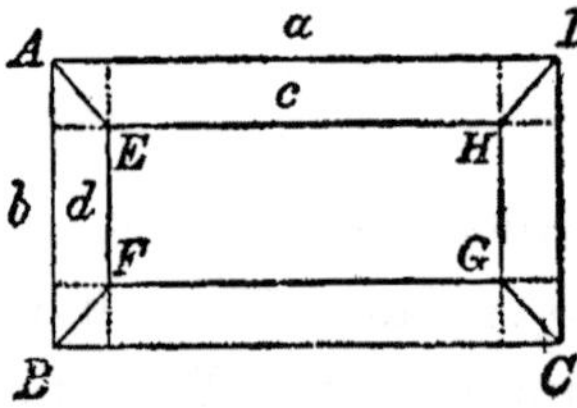

$$=z\{cd+\tfrac{1}{2}(a-c)d+\tfrac{1}{2}(b-d)c+\tfrac{1}{3}(a-c)(b-d)\}$$
$$=z\times\tfrac{1}{6}\{ab+cd+(a+c)(b+d)\}.$$

The last expression stated in words leads to the rule. The last part of the rule relating to the volumes of pyramids and cones is well known. Ganesa and Súryadása give curious demonstrations of the rule. See the foot-notes.]

---

[1] As the bottom of the acute excavation is deep, by finding an area for it in the manner before directed, the regular equal solid is produced; wherefore proportion is made: if such be the content, assuming three places, what is the content taking one? Thus the content of the regular equal solid, divided by three, is that of the acute one.—Súr.

222. Example. Tell the quantity of the excavation in a tank, of which the length and breadth are equal to twelve and ten cubits at its mouth, and half as much at the bottom, and of which the depth, friend, is seven cubits.

Statement : length 12 ; breadth 10 ; depth 7. The area at the mouth is 120 ; at the bottom 30 ; reckoned by the sum of the sides 270. Total 420. Mean area 70. Solid content 490.

223. Example. In a quadrangular excavation, the side being equal to twelve cubits, what is the content, if the depth be measured by nine ? and in a round one, of which the diameter is ten and depth five ? and tell me separately, friend, the content of both acute solids.

Statement of quadrangular tank : side 12 : depth 9. Proceeding as directed, the solid content comes out 1296. The content of the acute solid (quadrangular pyramid) is 432.

Statement of round tank : diameter 10 ; depth 5. The content nearly exact is $\frac{3927}{10}$; of the acute solid (cone), $\frac{1309}{10}$. Or gross content of the cylindrical tank is $\frac{2750}{7}$; of the cone, $\frac{2750}{21}$.

[The value of π is taken to be $\frac{3927}{1250}$. (See § 201.)]

CHAPTER VIII.

STACKS.[1]

224—225. Rule[2] : a stanza and a half. The area of the plane figure (or base) of the stack,[3] multiplied by the height,[4] will be the solid content. The content of the whole pile, being divided by that of one brick, the number of bricks is found. The height of the stack, being divided by that of one brick, gives the number of layers.[5] So likewise with piles of stones.

[The stack is supposed to be in the form of a rectangular parallelopiped, and the reason for the rule is obvious. Bricks are, however, usually arranged in a pile so as to form a frustum of a quadrangular pyramid.]

226—227. Example : two stanzas. The bricks of a pile being eighteen fingers long, twelve broad and three high, and the stack being five cuibts broad, eight long and three high, say what the solid content of the pile is ; and what the number of bricks, and how many the layers.

---

1 *Chiti-vyavahára.*

2 To find the solid content of a pile of bricks, or of stones or other things of uniform dimensions : also the number of bricks and of strata contained in the stack.

3 *Chiti*, a pile or stack.

4 *Uchchhráya*, *uchchhriti*, *auchchya*, height.

5 *Stara*, layer or stratum.

Statement : length of pile, 8 ; breadth, 5 ; height, 3. Bricks, $\frac{3}{4}$ by $\frac{1}{2}$ by $\frac{1}{8}$.

Answer. Solid content of the brick, $\frac{3}{64}$; of the stack, 120. Number of bricks, 2560. Number of layers, 24.

So likewise in the case of a pile of stones.

CHAPTER IX.

SAW.[1]

228. Rule: two half stanzas.[2] Half the sum of the thickness at both extremities, multiplied by the length in fingers, and the product again multiplied by the number of sections of the timber, and divided by five hundred and seventy-six,[3] will be the measure in cubits.

[The faces of the timber to which the sections are parallel, are supposed to be trapeziums, and the ends are supposed to be rectangular. Let $ABCD$ represent one of the sections, Then

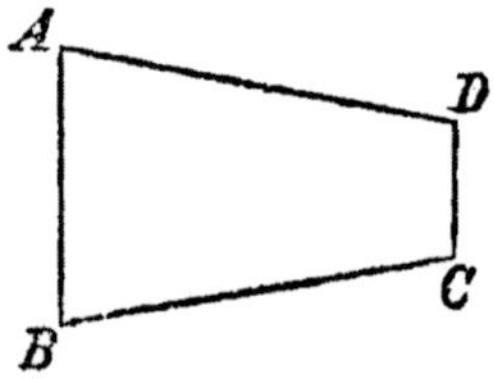

its area $=\frac{1}{2}(AB+CD)\times$ perpendicular distance between AB and $CD=\frac{1}{2}(AB+CD)\times AD$, nearly. The object of the reckoning is to settle the sawyer's charge which is at a certain rate for each square cubit along which the sawing is made. Hence the above area must be multiplied by the number of sections to get the total area for which the charge is to be reckoned.]

229. Example. Tell me quickly, friend, what the reckoning will be in cubits, for a timber the thickness of which is twenty fingers at the root, and sixteen

---

[1] *Krakacha-vyavahára*, determination of the reckoning concerning the saw (*krakacha*) or iron instrument with a jagged edge for cutting wood.—Súr.

[2] The concluding half of one stanza, begun in the preceding rule (§ 225), and the first half of another stanza of like metre completed in the following rule (§ 230).

[3] To reduce superficial fingers to superficial cubits.

fingers at the tip, and the length a hundred fingers, and which is cut by four sections.

Statement : length 100 ; thicknesses 20 and 16. Number of sections, 4.

Half the sum of the thicknesses at the two extremities, 18, multiplied by the length, makes 1800 ; this multiplied by the number of sections, gives 7200 ; divided by 576, gives the quotient in cubits, $\frac{25}{2}$.

230. Rule : half a stanza. But when the wood is cut across, the superficial measure is found by the multiplication of the thickness and breadth, in the mode above mentioned.[1]

[The reason for the rule is obvious.]

231. Maxim. The price for the stack of bricks or the pile of stones, or for excavation and sawing, is settled by the agreement of the workman, according to the softness or hardness of the materials.[2]

232. Example. Tell me what the superficial measure in cubits will be, for nine cross sections of a timber, of which the breadth is thirty-two fingers, and thickness sixteen.

Statement: breadth 32 ; thickness 16. Number of sections, 9.

Answer : 8 superficial cubits.

[The timber is supposed to be in the form of a rectangular parallelopiped.]

---

[1] If the breadth be unequal, the mean breadth must be taken.—Gan. and Súr.

[2] This is levelled at certain preceding writers who have given rules for computing specific prices or wages, as A'rya-bhatta quoted by Ganesa, and as Brahmagupta (XII, 49) ; particularly in the instance of sawyer's work, by varying the divisors according to the difference of the timber.

CHAPTER X.

MOUND[1] OF GRAIN.

233. Rule. The tenth part of the circumference is equal to the depth (height[2]) in the case of coarse grain; the eleventh part, in that of fine; and the ninth, in the instance of bearded corn.[3] A sixth of the circumference being squared and multiplied by the depth (height), the product will be the solid cubits:[4] and they are *khárís* of *Magadha.*[5]

---

[1] *Rási-vyavahára*, determination of a mound (of grain).

[2] *Bedha*, depth. Here it is the height in the middle from the ground to the summit of the mound.—Súr.

[3] *Anu, súkshma-dhánya*, fine grain, as mustard seed, &c.—Gan. As Paspalum Kora, &c.—*Mano.* As wheat, &c.—Súr.

Ananu, sthúla-dhánya, coarse grain, as chiches (cicer arietinum).—Gan. and Súr. As wheat &c.—*Mano.* Barley, &c.—Chaturveda on Brahm.

Súkin, súka-dhánya, bearded corn, as rice, &c.

The coarser the grain, the higher the mound. The rule is founded on trial and experience; and for other sorts of grain, other proportions may be taken, as 9½ or 10½ or 12 times the height, equal to the circumference.—Gan. and Súr. The rule is taken from Brahmagupta, XII, 50.

[4] This is a rough calculation, in which the diameter is taken at one-third of the circumference. The content may be found with greater precision by taking a more nearly correct proportion between the circumference and diameter.—Gan.

[5] See § 7. The proportion of the *khárí* or other dry measure of any province to the solid cubit being determined, a rule may be readily formed for computing the number of such measures in a conical mound of grain, Ganesa accordingly delivers rules by him devised for the *khárí* of *Nandigráma* and for that of *Devagiri*: 'the circumference measured by the human cubit, squared and divided by sixteen, gives the *khárí* of *Nandigráma*; and by sixty, that of *Devagiri*.' (*Devagiri*, lit. mountain of the

[The mound is supposed to be conical, the height being stated arbitrarily. The circumference of the base being given, the height will of course depend on the vertical angle of the cone. The rule is very rough, the value of π being taken equal to 3, as Ganesa remarks.

Let r denote the radius of the base, and h the height. Then the volume of the mound $= \frac{1}{3}\pi r^2 h = \frac{(2\pi r)^2}{3 \times 4\pi} \times h = \left(\frac{\text{circumference}}{6}\right)^2 . h$, supposing $\pi = 3$.]

234. Example. Mathematician, tell me quickly how many *khárís* are contained in a mound of coarse grain standing on even ground, the circumference of which (mound) measures sixty cubits; and separately say how many (are there) in a like mound of fine grain and in one of bearded corn.

Statement: circumference 60; height 6

Answer: 600 *khárís* of coarse grain. But of fine grain, the height is $\frac{60}{11}$, and quantity thence deduced, $\frac{6000}{11}$ So, of bearded corn, the height is $\frac{60}{9}$, and quantity $\frac{6000}{9}$ *khárís*.

235. Rule. In the case of a mound piled against the side of a wall, or against the inside or outside of a corner of it, the product is to be sought with the circumference multiplied by two,

---

gods, is better known by the name of *Daulatabad*, which the Emperor Muhammad conferred on it in the 14th century. *Nandigráma*, lit. the town or village of *Nandi*, Siva's bull and vehicle, retains the antique name, and is situated about 65 miles west of *Devagiri*.) He further observes that the cubit intended by the text is a measure in use with artisans, called in vulgar speech *gaj*; and a *khárí* equal to such a solid cubit will contain twenty-five *manas* and three quarters.

four, and one and a third ; and is to be divided by its own multiplier.[1]

[The reason for the rule is clear from the foot-note. The circumferences that are supposed to be given in the three cases are respectively half, one-fourth, and three-fourths of that of the base of the complete conical mound.]

236—237. Example: two stanzas. Tell me promptly, friend, the number of solid cubits contained in a mound of grain, which rests against the side of a wall, and the circumference of which measures thirty cubits ; and that contained in one piled in the inner corner and measuring fifteen cubits ; as also in one raised against the outer corner and measuring nine times five cubits.

Statement :

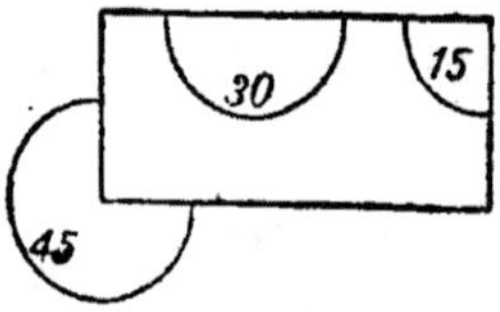

Twice the first mentioned circumference is 60. Four times the next is 60. The last multiplied by one and a third is likewise 60. With these the product is alike 600. This being divided by the respective multipliers, gives the several answers,[2] 300, 150 and 450.

[1] Against the wall, the mound is half a cone ; in the inner corner, a quarter of a cone ; and against the outer corner, three quarters. The circumference intended is a like portion of a circular base ; and the rule finds the content of a complete cone, and then divides it in the proportion of the part. See Gan., &c.

[2] For coarse grain : but the product is $\frac{6000}{11}$ for fine, and $\frac{6000}{9}$ for bearded corn ; and the answers are $\frac{3000}{11}$, $\frac{1500}{11}$, $\frac{4500}{11}$; and $\frac{3000}{9}$, $\frac{1500}{9}$, $\frac{4500}{9}$,—Gan. &c.

CHAPTER XI.

SHADOW[1] OF A GNOMON.

238. Rule.[2] The number five hundred and seventy-six being divided by the difference of the squares of the differences of both shadows and of the two hypotenuses,[3] and the quotient being added to one, the difference of the hypotenuses is multiplied by the square root of that sum; and the product being added to, and subtracted from, the difference of the shadows, the moieties of the sum and difference are the shadows.

[The translation of the last sentence is not quite correct. It should be, "and the difference of the shadows being added to and subtracted from the product, the moieties, &c."

The rule, as the author hints in the example which follows (§239), is founded on the algebraic solution of a quadratic equation. Ganesa gives it at length after the manner of the author's *Víja-ganita*. It is however very

---

[1] *Chháyá-vyavahára*, determination of shadow; that is measurement by means of a gnomon:

[2] The difference of the shadows and difference of the hypotenuses being given, to find the length of the shadows and hypotenuses—Súr,

This rule is the first in the chapter, according to all the commentators except Súryadása, who begins with the next, §240, and places this after §244.

[3] *Chháyá*, *bhá*, *prabhá*, shadow.

Sanku, *nara*, *nri*, a gnomon, usually 12 fingers long.

Karna, hypotenuse of the triangle, of which the gnomon is the perpendicular, and the shadow the base.

long and not at all clear, and so it has not been given in the foot-note. The rule may be demonstrated after the manner of modern algebra as follows :—

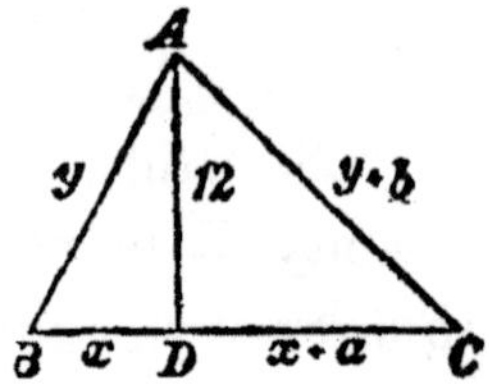

Let ABC be a triangle, and AD perpendicular to BC. Then BD, DC are called the shadows of the *sanku* or gnomon AD, which is supposed to be 12 fingers long ; BD, DC being supposed to be the shadows on a horizontal plane of the gnomon AD, produced either by sunlight at two different hours of the day, or by artificial light. The object of the rule is to find these shadows, the difference of AC, AB, and of CD, DB being supposed known.

Let $BD=x$, $DC=x+a$, $AB=y$, $AC=y+b$, a and b being known, and the measurements being in fingers.

Then $y^2-x^2=(y+b)^2-(x+a)^2=144$;

$$\therefore \quad by=ax+\frac{a^2-b^2}{2},$$

whence by squaring and substituting x^2+144 for y^2, we obtain the quadratic

$$x^2+ax+\left(\frac{a^2-b^2}{4}-\frac{144b^2}{a^2-b^2}\right)=0,$$

solving which we get $x=\frac{1}{2}\left\{-a+b\surd\left(1+\frac{576}{a^2-b^2}\right)\right\}$,

(the upper sign only being admissible),

and $x+a=\frac{1}{2}\left\{a+b\surd\left(1+\frac{576}{a^2-b^2}\right)\right\}$.

These results stated in words lead to the rule. The rule is not of much importance.]

239. Example. The ingenious man, who tells the shadows of which the difference is measured by nineteen, and the difference of hypotenuses by

thirteen, I take to be thoroughly acquainted with the whole of algebra as well as arithmetic.

Statement: difference of shadows, 19; difference of hypotenuses, 13. (Gnomon 12.)

Difference of their squares 192. By this divide 576: quotient 3. Add one. Sum 4. Square root 2. By this multiply the difference of hypotenuses 13: product 26. Add it to, and subtract it from, the difference of the shadows 19.

[The translation here is incorrect; it should be, "add to it, and subtract from it, the difference, &c." See note to §238.]

Half the sum and difference are the shadows, *viz.*, $\frac{45}{2}$ and $\frac{7}{2}$.

Under the rule in § 134, the gnomon being the upright, and the shadow the side, the square root of the sum of their squares is the hypotenuse. Thus the hypotenuses are $\frac{51}{2}$ and $\frac{25}{2}$.

240. Rule[1]; half a stanza. The gnomon multiplied by the distance of its foot from the foot of the light, and divided by the height of the torch's flame less the gnomon, will be the shadow.

---

[1] The elevation of the light and (horizontal) distance of its foot from the foot of the gnomon being given, to find the shadow.—Gan.

The rule follows from similar triangles as explained by Súryadása, as follows :—

Let A be the position of the light, CD the gnomon, and DE its shadow. From A draw AB perpendicular to ED produced. Through D draw DF parallel to AE. Then from the similar triangles CDE, FBD, we get $\frac{DE}{DC}=\frac{BD}{BF}$, whence

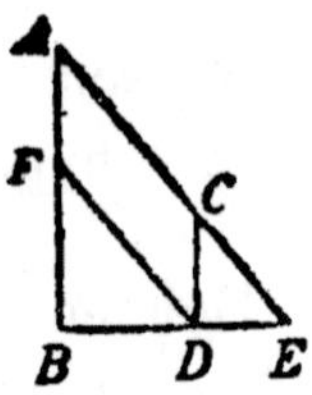

$DE=\frac{BD \cdot DC}{AB-CD}$, $\because$ $CD=AF$. Hence the rule.]

241. Example. If the base between the gnomon and torch be three cubits, and the elevation of the light, three cubits and a half, say quickly, friend, how much the shadow of a gnomon will be which measures twelve fingers.

Statement : gnomon $\frac{1}{2}$; distance between gnomon and torch, $\frac{7}{2}$; elevation of the light, 3.

Answer. Proceeding as directed, the shadow comes out 12 fingers.

242. Rule[1] : half a stanza. The gnomon being multiplied by the distance between it and the light, and divided by the shadow, and the quotient being added to the gnomon, the sum is the elevation of the torch.

[As Súryadása remarks, this rule also follows from similar triangles. See figure in the note to § 240.]

---

[1] To find the elevation of the torch, the length of the shadow, and the (horizontal) distance being given.—Súr.

243. Example. If the base between the torch and gnomon be three cubits, and the shadow be equal to sixteen fingers, how much will be the elevation of the torch? And tell me what the distance is between the torch and gnomon (if the elevation be given.)

Statement: distance between torch and gnomon, 3; shadow $\frac{2}{3}$.

Answer: height of the torch $\frac{11}{4}$.

244. Rule[1]: half a stanza. The shadow, multiplied by the elevation of the light less the gnomon and divided by the gnomon, will be the interval between the gnomon and light.

This like the preceding rule also follows from similar triangles.]

Example, as before proposed (§ 243.)

Answer: distance 3 cubits.

245. Rule[2]: a stanza and a half. The length of a shadow multiplied by the distance between the terminations of the shadows and divided by the difference of the lengths of the shadows, will be the base. The product of the base and the gnomon, divided by the length of the shadow, gives the elevation of the torch's flame.[3]

---

[1] To find the (horizontal) distance, the elevation of the torch and length of the shadow being given.—Gan. and Súr.

[2] The gnomon being set up successively in two places, the distance between which is known, and the length of the two shadows being given, to find the elevation of the light, and the base.—Gan. and Súr.

[3] The rule is borrowed from Brahmagupta (XII, 54).

In like manner is all this, which has been before declared, pervaded by the Rule of Three with its variations, as the universe is by the Deity.[1]

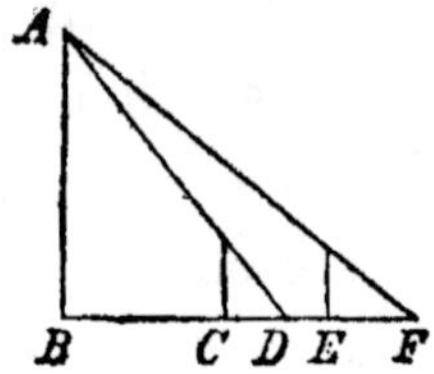

[Let A be the position of the light, C, E, the positions of the foot of the gnomon, and CD, EF, the corresponding shadows. Let $BC=x$, $BA=y$, $CD=a$, $EF=b$, $CE=c$, the measurements being in fingers.

Then from similar triangles we have,

$$\frac{y}{x+a}=\frac{12}{a},\quad \frac{y}{x+b+c}=\frac{12}{b};$$

whence $x=\dfrac{ac}{b-a}$.

$$\therefore\quad x+a=\frac{a(b+c-a)}{b-a}\text{.........(1)},$$

and $$y=\frac{12\,(x+a)}{a}=\frac{12\,(x+b+c)}{b},$$

whence the reason for the rule is obvious. Bháskara's own explanation is practically the same as the above, but it is not clearly put. He at once states a proportion which is equivalent to equation (1) above, but he does not explain how it is obtained.]

246. Example. The shadow of a gnomon measuring twelve fingers being found to be eight, and that of the same placed on a spot two cubits further in the same direction, being measured twelve fingers, say, intelligent mathematician,

---

[1] The author intimates that the whole preceding system of computation, as well as the rules contained under the present head, as those before delivered, is founded on the rule of proportion.—Gan.

how much the distance of the shadow[1] from the torch is, and the height of the light, if thou be conversant with computation, as it is termed, of shadow.

Statement : shadows, 8, 12 ; interval between the positions of the foot of gnomon, 48.

Here the interval between the termination of the shadows is in fingers 52. The first shadow 8, multiplied by the interval 52, and divided by the difference of the length of the shadows, *viz.*, 4, gives the length of the base 104. It is the distance between the foot of the torch and the tip of the first shadow. So the length of the base to the tip of the second shadow is 156.

The product of the base and gnomon, divided by the shadow, gives both ways the same elevation of the light, *viz.*, 6½ cubits.

"In like manner, &c."[2]—Under the present head of measurement of shadow, the solution is obtained by putting a proportion : *viz.*, if so much of the shadow, as is the excess of the second above the first, give the base intercepted between the tips of the shadows, what will the first give ? The distances of the terminations of the shadows from the foot of the torch are in

---

[1] All the commentators appear to have read 'gnomon' in this place ; but one copy of the text exhibits 'shadow' as the reading : and this seems to be correct.

[2] Reference to the text, §245. [The author here purports to explain fully what he has hinted at before.—Ed.]

this manner severally found. Then a second proportion is put : if, the shadow being the side, the gnomon be the upright; then, the base being the side, what will be the upright? The elevation of the torch is thus found; and is both ways (that is, computed with either shadow,) alike.

[See note to § 245.]

So the whole sets of five or more terms are explained by twice putting three terms and so forth.

As the Being, who relieves the minds of his worshippers from suffering, and who is the sole cause of the production of this universe, pervades the whole, and does so with his various manifestations, as worlds, paradises,[1] mountains, rivers, gods, demons, men, trees,[2] and cities; so is all this collection of instructions for computations pervaded by the rule of three terms. Then why has it been set forth by so many different (writers,[3] with much labour and at great length)? The answer is :—

247. Whatever is computed either in algebra or in this (arithmetic) by means of a multiplier and

---

[1] *Bhuvana*, worlds, *Bhavana*, paradises, the abodes of Brahmá and the rest of the gods. [The reading here adopted by Colebrooke is apparently different from that in Pandit Jívánanda Vidyásságara's edition in which we have *sakala-bhuvana-bhávanesu*, which rendered becomes, 'who is the creator of all the worlds,' the words, 'world's and 'paradises' in Colebrooke's translation, being omitted in that case.—Ed.]

[2] *Naga*, either tree or mountain. The term, however, is read in the text by none of the commentators besides Ganesa.

[3] As Srídhara and the rest.—*Mano.* As Brahmagupta and others.—Gang.

a divisor, may be comprehended by the sagacious learned as the rule of three terms. Yet has it been composed by wise instructors in miscellaneous and other manifold rules, teaching its easy variations, thinking thereby to increase the intelligence of such dull comprehensions as ours.

[The following example occurs in the *Vija-gunita* :—

The shadow of a gnomon twelve fingers high being lessened by a third part of the hypotenuse, became fourteen fingers long. Tell quickly, mathematician, that shadow.†

Let x denote the length of the shadow. Then the hypotenuse $=\sqrt{x^2+144}$, and we get the equation, $x-\frac{1}{3}\sqrt{x^2+144}=14$, whence $2x^2-63x+405=0$, or $(2x-45)(x-9)=0$; $\therefore\ x=\frac{45}{2}$ or 9. The second value being less than 14 is inadmissible as is noticed by Bháskara.

---

† कर्णस्य त्रिलवेनोना द्वादशाङ्गुलशङ्कुभा ।
चतुर्दशाङ्गुला जाता गणक ब्रूहि तां द्रुतम् ॥

CHAPTER XII.

PULVERIZER.

248—252. Rule : five stanzas.

248. In the first place, as preparatory to the investigation of a pulverizer,[1] the dividend, divisor

---

[1] *Kuttaka-vyavahára* or *kuttakádhyáya*, determination of a grinding or pulverizing multiplier, or quantity such that a given number being multiplied by it, and the product added to a given quantity, the sum (or, if the additive be negative, the difference) may be divisible by a given divisor without remainder. *Kuttaka* or *kutta* from *kutt*, to grind or pulverize ; (to multiply : all verbs importing tendency to destruction also signifying multiplication.—Gan.) The derivative import of the word is retained in the present version to distinguish it from multiplier in general ; *kuttaka* being restricted to the particular multiplier of the problem in question.

According to the remark of Ganesa, his chapter as well as the following chapter on combination belongs to algebra rather than arithmetic : and they are here introduced, as he observes, and treated without employing algebraic forms, to gratify such as are unacquainted with analysis. See *Vija-ganita*. Chap. II, from which the present chapter is borrowed, the contents being copied, with some variation of the order, nearly word for word.

Ganesa notices an objection, namely, that this subject ought not to have been introduced into a treatise on arithmetic, while a passage of A'ryabhatta expressly distinguishes it from both arithmetic and algebra : "the multifarious doctrine of the planets, arithmetic, the pulverizer (*kuttaka*), and analysis (*víja*) and the rest of the science treating of seen (or physical) objects." He answers the objection by saying that mathematics (*ganita*) consists of two branches treating of known and of unknown quantity (*vyakta-ganita* and *avyakta-ganita*) ; that the investigation of the pulverizer is comprehended in algebra ; and that the separate mention of this subject by A'ryabhatta and other ancient authors is intended to

and additive quantity[1] are, if practicable, to be reduced by some number.[2] If the number, by which the dividend and divisor are both measured, do not also measure the additive quantity, the question is an ill put (or impossible) one.

249—251. The last remainder, when the dividend and divisor are mutually divided, is their common measure.[2] Being divided by that common measure, they are termed reduced quantities.[3] Divide mutually the reduced dividend and divisor, until unity be the remainder in the dividend. Place the quotients one under the other, and the additive quantity beneath them, and cipher at the bottom. By the penult multiply the number next above it and add the lowest term. Then reject the last and repeat the operation until a pair of numbers be left. The uppermost of these being abraded[4] by the reduced dividend, the remainder

---

indicate its difficulty and importance. In Brahmagupta's work, the whole of algebra is comprised under the title of *kuttakádhyáya*, chapter on the pulverizer. See Brahm., Ch. XVIII.

[1] *Kshepa* or *yuti*, additive ; from *kship* to cast or throw in, and from *yu* to mix. *Visuddhi*, subtractive quantity.

[2] *Apavartana*, abridgment.—Gan. Reduction to least terms, division without remainder ; also the number which serves to divide without residue, the common measure.—Súr. [It is really the greatest common measure.—Ed]

[3] *Dridha*, firm ; reduced by the common divisor to the least term. The word is applicable to the reduced additive, as well as to the dividend and divisor.

[4] *Tashta*, abraded ; from *taksh*, to pare or abrade : divided, but the residue taken, disregarding the quotient.—Súr. As it were a residue after repeated subtractions.—Gang.

Takshana, the abrador : the divisor employed in such operation

is the quotient. The other (or lowermost) being in like manner abraded by the reduced divisor, the remainder is the multiplier.

252. Thus precisely is the operation when the number of quotients is even. But if the number be odd, the numbers as found must be subtracted from their respective abraders, and the residues will be the true quotient and multiplier.

[This Chapter, as Ganesa remarks, properly belongs to Algebra and not to Arithmetic. We have already seen, however, that the present treatise deals with both Arithmetic and Algebra.

The whole of this Chapter is occupied with problems producing indeterminate equations of the first degree, and the object of the several rules is to find positive integral solutions of such equations. The reason for the above rule will be best understood from the example in § 253. Let y denote the multiplier and x the integral quotient. Then we get $\frac{221y+65}{195}=x$, and the object of the rule is to find positive integral solutions of this equation.

Dividing by the common measure 13, we get $\frac{17y+5}{15}=x$,

$\therefore \quad 15x-17y=5$........(1).

This equation is of the type $Ax-By=C$, A being less than B, and the rule refers to cases of this class. Convert $\frac{B}{A}$ into a continued fraction, and suppose the result is $\frac{B}{A}=a+\frac{1}{b+c}\,\frac{1}{+\&c}$. Let $\frac{q}{p}$ be the convergent immediately preceding $\frac{B}{A}$. Then we know that $x=qC$, $y=pC$, or $x=(B-q)\,C$, $y=(A-p)\,C$, is one solution of equation (1), according as $Aq-Bp=\pm\ 1$; and that the general solution is $x=\alpha+Bt$, $y=\beta+At$, where α and β are one set of values of x and y, and where we may give to t any positive integral value, and also such negative integral values as make Bt and At numerically less than α and β respectively. (See Todhunter's Algebra, Arts. 630, 631.) Now the successive convergents to $\frac{B}{A}$ are $\frac{a}{1}$, $\frac{ab+1}{b}$, $\frac{abc+a+c}{bc+1}$, &c., the law of formation being well known (See Todhunter's Algebra, Art. 604); and the object of the rule in § 251 is to find the value of the convergent $\frac{q}{p}$, and thence the value of the quantities qC, pC. The rule, however, is not clearly expressed, and it is difficult to understand its working. An explanation of the rule by means of a particular example is given by Krishna in his commentary on the *Vija-ganita*, from which we may deduce the following general explanation. Let us first consider the case where the number of quotients exclusive of the last one is two, *viz.*, a, b, the additive being C. Then according to the rule we get the series a, b, C, 0. The rule next directs us to multiply the penultimate (C) by the number preceding it, *viz.*, b, and to add the last term, *viz.*, 0, to the product. We thus get $bC+0$ or bC, and we have now to replace the previous multiplier b by this quantity bC, and reject the last term 0. We thus have the new series, a, bC, C, with which we have to repeat the above operation : that is, we have to multiply the penultimate bC, by a, and add the last term

C to the product; whence we get $abC+C$, by which we have to replace the multiplier a, and we have to reject C. The series thus becomes $abC+C$, bC, and this consisting only of two terms, we infer according to the rule that $qC=abC+C$, and $pC=bC$. But we know as a matter of fact that in the case we are considering, $q=ab+1$, and $p=b$; thus the rule holds good in this case. We may now prove by induction that the rule holds universally. For supposing the number of quotients exclusive of the last one to be three, *viz.*, a, b, c, all we have got to do is to take the additional quotient c, so that we have to write $b+\frac{1}{c}$ for b in the above expressions for qC, pC; and we get in this case $\frac{qC}{pC}=\frac{a\left(b+\frac{1}{c}\right)C+C}{\left(b+\frac{1}{c}\right)C}=\frac{a(bcC+C)+cC}{bcC+C}$,

so that $qC=a(bcC+C)+cC$, and $pC=bcC+C$. But it is easy to see from the very nature of these expressions that they are precisely what we would get by means of the rule, if we take into consideration the quotient c. Thus we see that the rule is universally true. In the present example, $a=1$, $b=7$, $c=2$, and 7 is the last quotient but one. Hence, $q=ab+1=7+1$, and $qC=7\times5+5$; $p=b=7$, and $pC=7\times5$. And $\because$ in this case $Aq-Bp=1$, $\therefore$ $x=40$, $y=35$, is one solution. Hence putting $\alpha=40$, $\beta=35$, and $t=-2$, in the general expressions for x and y, we get $x=40-17\times2=6$, $y=35-15\times2=5$. These are the least positive integral values. Putting $t=-1$, we get $x=23$, $y=20$, and so on. Or taking $x=6$, $y=5$, as one solution, we may get others from the expressions $6+17\,t$, $5+15\,t$, by giving to t any positive integral value. Thus putting $t=1$, we get $x=23$, $y=20$.

putting $t=2$, we get $x=40$, $y=35$; and so on. Thus the reason for the process in § 253, as well as that for the rule in § 262, is clear. The meaning of the term *abraded* as used by the author is also clear.]

253. Example. Say quickly, mathematician, what that multiplier is, by which two hundred and twenty-one being multiplied, and sixty-five added to the product, the sum divided by a hundred and ninety-five becomes exhausted.

Statement: dividend 221
divisor 195. Additive 65.

Here the dividend and divisor being mutually divided, the last of the remainders (or divisors) is 13. By this common measure, the dividend, divisor and additive, being reduced to their least terms, are dividend 17, divisor 15, additive 5. The reduced dividend and divisor being divided reciprocally, and the quotients put one under the other, the additive under them, and cipher at the bottom, the series[1] which results is 1.
7
5
0

Then multiplying by the penult the number above it and proceeding as directed, the two quantities are obtained 40. These being abraded by
35

[1] *Valli*, series.

the reduced dividend and divisor 17 and 15, the quotient and multiplier are obtained 6 and 5. Or, by the subsequent rule (§ 262), adding them to their abraders multiplied by an assumed number, the quotient and multiplier (putting 1) are 23 and 20; or putting 2, they are 40 and 35; and so forth.

254. Rule. The multiplier is also found by the method of the pulverizer, the additive quantity and dividend, being either reduced by a common measure (or used unreduced[1]). But if the additive and divisor be so reduced, the multiplier found, being multiplied by the common measure, is the true one.

[The reason for the rule will appear from the solution of the example in § 255.]

255. Example. If thou be expert in the investigation of such questions, tell me the precise multiplier by which a hundred being multiplied, with ninety added to the product, or subtracted from it,[2] the sum or difference may be divisible by sixty-three without a remainder.

Statement : dividend 100, divisor 63. Additive 90.

The quotient and multiplier are found by proceeding as before, 30 and 18.

---

[1] Gan.

[2] An example of the subsequent rule in § 256.

Or, the dividend and additive being reduced by the common measure ten, we get dividend 10, divisor 63, additive 9. Placing the quotients of reciprocal division, the additive quantity and cipher, one under the other, the series is $\begin{array}{c}0\\6\\3\\9\\0\end{array}$. And the multiplier is found by the former process 45. The quotient[1] (3) is here not to be taken; and the number of quotients (of the series) being odd, the multiplier 45 is to be subtracted from its own abrader 63; the true multiplier is thus found 18. The dividend being multiplied by that multiplier, and the additive quantity being added, and the sum divided by the divisor, the quotient is found 30.

Or, the divisor and additive quantity being reduced by the common measure nine, we get dividend 100, divisor 7, additive 10. Here the quotients, the additive and cipher make the series $\begin{array}{c}14\\3\\10\\0\end{array}$. The multiplier is found 2, which multiplied by the common measure 9, gives the true multiplier 18.

---

[1] [This probably means the *last* quotient. There is, however, no force in the remark; the last quotient being always excluded under the rule in §§ 250–251.—ED.]

Or, the dividend and additive being reduced, and further the divisor and additive, by common measures, we get dividend 10, divisor 7, additive 1. Proceeding as before the series is 1.
2
1
0

The multiplier hence deduced is 2, which taken into the common measure 9, gives 18; and hence, by multiplication and division, the qnotient comes out 30.

Or, adding the quotient and multiplier as found, to (multiples of) their respective divisors, the quotient and multiplier are 130 and 81; or 230 and 144; and so forth.

[Putting y = multiplier, and x = quotient, we get the equation $63x-100y=90$......(1). If we convert $\frac{100}{63}$ into a continued fraction, the number of quotients will be found to be large, and so it will be tedious to form $\frac{q}{p}$. To make the process shorter, let us first consider the equation $63x-10y=9$......(2). Now if from (2) we find sets of positive integral values of x and y, it is clear that those values of y and 10 times the corresponding values of x will be sets of positive integral values of x and y satisfying (1). The general solution of (2) will be found to be $x=(10-3)9+10t$, $y=(63-19)9+63t$. Putting $t=-6$, we get $x=3$, $y=18$, as the least values. Thus $x=30$, $y=18$, are the least values satisfying (1). In the text, the least value of y is found from the expression

$63-(19\times9+63t')$, by putting $t'=-2$, and the reason for this is stated to be the fact that the number of quotients (exclusive of the last one) is odd. The explanation is that the number of quotients exclusive of the last one being odd, $\frac{q}{p}$ is less than $\frac{B}{A}$ (Todhunter's Algebra, Art. 603), and so $Aq-Bp=-1$, and $\therefore$ the general solution is $x=(B-q)C+Bt$, $y=(A-p)C+At$......(3). In the text, the value of y is derived from the expression $A-(pC+At')$...(4). Supposing $\therefore$ that the expressions in (3) and (4) give the same values of y for certain values of t and t', the relation between such values will be given by $t+t'=1-C$. Thus it is clear that we can derive values of y from both these expressions, but not from the expression $pC+At$, when the number of quotients exclusive of the last one is odd This also explains the statement in §252.

Similarly, if we find values of x and y from the equation $7x-100y=10$, these values of x and 9 times the corresponding values of y will satisfy (1).

Lastly if we find values of x and y from the equation $7x-10y=1$, it is easy to see that 10 times these values of x, and 9 times the corresponding values of y will satisfy (1). Thus the reason for the rule in § 254 is obvious.]

256. Rule[1] : half a stanza. The multiplier and quotient, as found for an additive quantity, being subtracted from their respective abraders, answer for the same as a subtractive quantity.

[1] The rule serves when the additive quantity is negative.—Gan.and Súr.

Here the quotient and multiplier as found for the additive quantity ninety in the preceding example, namely, 30 and 18, being subtracted from their respective abraders, namely, 100 and 63, the remainders are the quotient and multiplier which answer when ninety is subtractive: *viz.*, 70 and 45.

Or, these being added to arbitrary multiples of their respective abraders, the quotient and multiplier are 170 and 108, or 270 and 171, &c.

[Let C be the additive or subtractive quantity. Then the corresponding equations will be $Ax-By= \ C$......(1).

$$Ax-By=-C......(2).$$

Let $x=\alpha$, $y=\beta$, be a solution of (1). Then $A\alpha-B\beta=C$.

$\therefore A(B-\alpha)-B(A-\beta)=-C$.

$\therefore x=B-\alpha$, $y=A-\beta$, is a solution of (2), whence the reason for the rule is clear. It will be readily seen that the general solution of (2) is $x=(B-\alpha)+Bt, y=(A-\beta)+At$.]

257. Another example[1]. Tell me, mathematician, the multipliers severally, by which sixty being mltiplied, and sixteen being added to the product, or subtracted from it, the sum or difference may be divisible by thirteen without a remainder.

Statement: dividend 60, divisor 13. Additive 16.

[1] This additional example is unnoticed by Ganesa, but expounded by the rest of the commentators, and found in all copies of the text that have been collated.

The series found as before, is 4.
1
1
1
1
16
0

Hence the multiplier and quotient are deduced 2 and 8. But the number of quotients (of the series) is here uneven ; wherefore the multiplier and quotient must be subtracted from their abraders 13 and 60 ; and the multiplier and quotient, answering to the additive quantity sixteen, are 11 and 52. These being subtracted from the abraders, the multiplier and quotient, corresponding to the subtractive quantity sixteen, are 2 and 8.

258. Rule[1] : a stanza and a half. The intelligent calculator should take a like quotient (of both divisions) in the abrading of the numbers for the multiplier and quotient (sought). But the multiplier and quotient may be found as before, the additive quantity being (first) abraded by the divisor ; the quotient, however, must have added to it the quotient obtained in the abrading of the additive. But in the case of a subtractive quantity, it is subtracted.

[The reason for the rule will appear from the solution of the example which follows.]

---

[1] Applicable when the additive quantity exceeds the dividend and divisor. —Gan.

259. Example. What is the multiplier, by which five being multiplied and twenty-three added to the product, or subtracted from it, the sum or difference may be divided by three without remainder ?

Statement : dividend 5 / divisor 3 . Additive 23.

Here the series is 1 and the pair of numbers found

1
23
0

as before 46 / 23 . They are abraded by the dividend and divisor, respectively. The lower number being abraded by 3, the quotient is 7 (and residue 2) The upper number being abraded by 5, the quotient would be 9 (and residue 1) ; nine, however, is not to be taken ; but, under the rule for taking like quotients, seven only, (and the residue censequently is 11). Thus the multiplier and quotient come out 2 and 11.

And by the former rule (§256) the multiplier and quotient answering to the same as a negative quantity are found, 1 and the negative quantity—6. Added to arbitrary multiples of their abraders, double for example, so that the quotient may be positive, the multiplier and quotient are 7 and 4. So in every (similar) case.

Or, Statement for the second (part of the) rule:

dividend 5
divisor 3 Additive abraded[1] 2.

The multiplier and quotient hence found as before are 2 and 4. These subtracted from their respective divisors, give 1 and 1, as answering to the subtractive quantity. The quotient obtained in the abrading of the additive, (*viz.* 7) being added in one instance and subtracted in the other, the results are 2 and 11 answering to the additive quantity, and 1 and -6 answering to the subtractive: or, to obtain a positive quotient, add to the latter twice their divisors: and the result is 7 and 4.

[Putting y=multiplier, and x=quotient, we get the equations $3x-5y=\pm 23$...(1). Taking the upper sign, the general solution will be found to be $x=2\times 23+5t$, $y=1\times 23+3t$. The least positive integral values are got by putting $t=-7$, *viz.*, $x=11$, $y=2$. The meaning of the first part of the rule in §258 is that the *same negative* value is to be given to t in the expressions for x and y. To explain the second part of the rule, we observe that equation (1) may be written $\frac{5y+2}{3}=x\mp 7=X$ suppose. We may then solve $3X-5y=2$......(2), the values of y being the same in (1) and (2), and the values of x being deducible from those of X, from the relation $x=X\pm 7$.]

260. Rule[2]: one stanza. If there be no additive

---

[1] 23, abraded by the divisor 3, gives the quotient 7 and residue 2.

[2] Applicable if there be no additive, or if it be divisible by the divisor without remainder.

quantity, or if the additive be measured by the divisor, the multiplier may be considered as cipher, and the quotient as the additive divided by the divisor.[1]

[The rule is clear enough.]

261. Example. Tell me promptly, mathematician, the multiplier by which five being multiplied and added to cipher, or added to sixty-five, the division by thirteen shall in both cases be without remainder.

Statement: dividend 5, divisor 13. Additive 0.

There being no additive, the multiplier and quotient are 0 and 0; or 13 and 5; or 26 and 10; and so forth.

Statement: dividend 5, divisor 13. Additive 65.

By the rule (§ 260), the multiplier and quotient come out 0 and 5; or 13 and 10; or 26 and 15; and so forth.

[Putting y=multiplier, and x=quotient, we get in the first case the equation $\frac{5y+0}{13}=x$; and in the second case, the equation $\frac{5y+65}{13}=x$. The general solution in positive integers of the first equation will be

[1] It is so in the latter case; but in the former (where the additive is null), the quotient is cipher.—*Súr.*

readily found to be $x=5r$, $y=13r$; and that of the second, $x=5(1+r)$, $y=13r$, where r may be zero or any positive integer.]

Rule[1]. Or, the dividend and additive being abraded by the divisor, the multiplier may thence be found as before; and the quotient from it, by multiplying the dividend, adding the additive, and dividing by the divisor.

In the former example (§ 253), the reduced dividend, divisor and additive respectively are, 17, 15, 5. Abraded by the divisor (15) the additive and dividend become 5 and 2; and the statement is:—

dividend 2
divisor 15   Additive 5.

Proceeding as before the two terms found are 5, 35. The second one, abraded by the divisor (15), gives the multiplier 5; whence, by multiplying with it the dividend (17) and adding (the additive), and dividing (by the divisor), the quotient comes out 6.

[The reason for the above rule is clear. Let the equation be $Ax-By=C$, and suppose B greater than A, and C less than A. Divide B by A; let K denote the quotient, and B' the remainder. Thus $B=KA+B'$, and $\therefore \frac{B}{A}=K+\frac{B'}{A}$. Convert $\frac{B}{A}$ and $\frac{B'}{A}$ into conti-

[1] This is found in one copy of the text, and is expounded only by Gangádhara, being unnoticed by the other commentators. It occurs, however, in the similar chapter of the *Víja-ganita*, § 62.

nued fractions, and let $\frac{q}{p}$ and $\frac{q'}{p'}$ be the convergents immediately preceding $\frac{B}{A}$ and $\frac{B'}{A}$. Then $\frac{q}{p}=K+\frac{q'}{p'}$ $=\frac{Kp'+q'}{p}$. Thus $p=p'$, and $Aq-Bp=A\ (Kp'+q')-(KA+B')p'=Aq'-B'p'$. Hence it is evident that the values of y found from $Ax-B'y=C$, will be the same as those found from $Ax-By=C$; and y being known, x is of course known from the equation $x=\frac{By+C}{A}$. It must be remarked here that the above rule applies only when the additive C is less than the divisor A, so that the additive abraded by the divisors remains unchanged.]

262. Rule for finding divers multipliers and quotients in every case: half a stanza. The multiplier and quotient, being added to their respective (abrading) divisors multiplied by assumed numbers, become manifold.

The influence and operation of this rule have been already shown in various instances.

[See note to § 252.]

263. Rule for a constant pulverizer[1]: one stanza. Unity being taken for the additive quantity, or for the subtractive, the multiplier and quotient, which may be thence deduced, being severally multiplied by an arbitrary additive or subtractive, and abraded by the respective divisors,

---

[1] *Sthira-kuttaka*, steady pulverizer.

will be the multiplier and quotient for such assumed quantity.

In the first example (§253), the reduced dividend and divisor with additive unity furnish this statement: dividend 17 / divisor 15. Additive 1.

Here the multiplier and quotient (found in the usual manner) are 7 and 8. These multiplied by an assumed additive five, and abraded by the respective divisors 15 and 17, give the multiplier and quotient 5 and 6, for that additive.

Next, unity being the subtractive quantity, the multiplier and quotient thence deduced are 8 and 9. These multiplied by five and abraded by the respective divisors, give 10 and 11.

So in every (similar) case.

Of this method of investigation great use is made in the computation of planets. On that account something is here said (by way of instance.)

[The above rule is not a new one. The equation is supposed to be $Ax-By=\pm 1$...(1), and from what we have already seen, the general solution of this is $x=q+Bt$, or $=(B-q)+Bt$, and $y=p+At$, or $=(A-p)+At$. If now the additive or subtractive be any integer whatever, *i.e.*, if the equation be $Ax-By=\pm C$...(2), we have only to multiply q or $B-q$, and p or $A-p$ by C in the above expressions for x and y, in order to get the general solution of equation (2). We may thus regard the general value of y found from (1) as a *steady* quantity from which we may derive the general value of y satisfying (2). This shows the propriety of the expression *constant pulverizer*.]

264. A stanza and a half. Let the remainder of seconds be made the subtractive quantity,[1] sixty the dividend, and terrestrial days[2] the divisor. The quotient deduced therefrom will be the seconds; and the multiplier will be the remainder of minutes. From this again the minutes and remainder of degrees are found; and so on upwards. In like manner, from the remainder of exceeding months and deficient days,[3] may be found the solar and lunar days.

The finding of (the place of) the planet and the elapsed days, from the remainder of seconds in the planet's place, is thus shown. Sixty is there made the dividend: terrestrial days, the divisor; and the remainder of seconds, the subtractive quantity:

[1] The present rule is for finding a planet's place and the elapsed time, when the fraction above seconds is alone given.—Gan.

[2] The number of terrestrial days in a *kalpa* is stated at 1577916450000. See the *Ganitádhyáya* of the *Siddhánta-siromani*, I, 20—21. [By a *terrestrial* day is meant the *mean solar* day, when it is taken for the purpose of astronomical measurement; but for practical purposes, it is taken as the time from sunrise to sunrise, which would make its duration variable. See *Goládhyáya*, Wilkinson's translation, II. 3, Bápú Deva Sástrí's note: *Súrya-siddhánta*, Burgess's translation, I, 34—40, note.—Ed.]

[3] *Adhi-mása*, additive months; *Avamadina*, subtractive days. See *Ganitádhyáya*, I, 42. [See also *Goládhyáya*, IV, 10—16, note: *Súrya-siddhánta*, I, 47—50, note.—Ed.] The *adhimásas* in a *kalpa* are 1593300000 [=1602999000000÷30−4320000000×12], being the excess of the lunar over the solar months. The *avamas* in a *kalpa* are 25082550000, being the excess of the lunar days over the terrestrial days.

with which the multiplier and quotient are to be found. The quotient will be seconds ; and the multiplier, the remainder of minutes. From this remainder of minutes taken (as the subtractive quantity), the quotient deduced will be minutes ; and the multiplier, the remainder of degrees. The residue of degrees is next the subtractive quantity ; terrestrial days, the divisor ; and thirty, the dividend : the quotient will be degrees ; and the multiplier, the remainder of signs. The twelve is made the dividend ; terrestrial days, the divisor ; and the remainder of signs the subtractive quantity : the quotient will be signs ; and the multiplier, the remainder of revolutions. Lastly, the revolutions in a *kalpa* become the dividend ; terrestrial days, the divisor ; and the remainder of revolutions, the subtractive quantity : the quotient will be the elapsed revolutions ; and the multiplier, the number of elapsed days.[1] Examples of this occur (in the *Siromani*) in the chapter of the problems[2] (*Triprasnádhyáya*).

In like manner, the exceeding months in a *kalpa* are made the dividend ; solar days,[3] the divisor ; and the remander of exceeding months, the sub-

---

[1] The elapsed days of the *kalpa* to the time for which the planet's place is found. See *Ganitádhyáya*, I, 47—49.

[2] [See also *Goládhyáya*, Chap. XIII.—Ed.]

[3] The solar days in a *kalpa* are 1555200000000 =[4320000000×360]. See *Ganitádhyáya*, I, 40.

[The number of solar years in a *kalpa* is 4320000000. See *Súrya-siddhánta*, I, 19. note.—Ed.]

tractive quantity : the quotient will be the elapsed additional months ; and the multiplier, the elapsed solar days. So the deficient days in a *kalpa* are made the dividend ; lunar days,[1] the divisor : and the remainder of deficient days, the subtractive quantity : the quotient will be the elapsed fewer days ; and the multiplier the elapsed lunar days.

[The reason for the rule for finding a planet's place and the elapsed time will be best understood from the illustration given by Ganesa and Gangádhara in arbitrary numbers. Put the terrestrial days in a *kalpa* 19, the revolutions of the planet in the *kalpa* 10, the elapsed days 12. Then we evidently get the proporton 19 : 12 : : 10 : number of revolutions already performed by the planet, whence the revolutions $=6\frac{6}{19}$. Thus the planet has performed 6 complete revolutions, and $\frac{6}{19}$ of a revolution, so that to find the planet's place, we must reduce the fraction $\frac{6}{19}$ to signs (*rásis*), degrees, minutes and seconds. Now as there are 12 signs in one revolution, 30 degrees in one sign, 60 minutes in one degree, and 60 seconds in one minute, we get $\frac{6}{19}$ of a revolution $=3$ signs, $23^\circ\ 41'.3''\ \frac{3}{19}$, and this result indicates the planet's place. Now suppose the remainder of seconds after division by 19, *i.e.*, 3, is alone given, and we have to find the planet's place by an inverse process. Let y denote the remainder of minutes, and x the integral number of seconds. Then it is clear from the process which we adopted in reducing

[1] The lunar days, reckoning thirty to the month or synodical revolution, are 1602999000000 in the *kalpa*. See *Ganitádhyáya*, I, 40,. [See also *Goládhyáya*, II, 3, note. The terrestrial days in a *kalpa* are 1577916450000. See *ibid.*, II, 3, note. These two numbers as given in the *Súrya-siddhánta* (I, 37) are slightly different.—Ed.]

the fraction $\frac{3}{19}$, that $\frac{60y-3}{19}=x$, the general solution of which is given (§256) by $x=60-(57+60t)$, $y=19$ $(18+19t)$. The only positive integral solution is got by putting $t=0$; then $x=3$, $y=1$. The quotient x is the number of seconds, *viz.*, 3; and the multiplier y is the remainder of minutes, *viz.*, 1. It is easy to see that there can be only one positive integral solution satisfying the conditions of the problem. For x must obviously be less than 60, and y less than 19; so that $57+60t$ must be positive and less than 60, and $18+19t$ must be positive and less than 19. Hence there can be only *one* value of t satisfying these conditions, and consequently only *one* positive integral solution satisfying the problem. Making the necessary changes in the coefficient of y and in the subtractive quantity, and repeating the above process, we clearly obtain the number of minutes, degrees and signs indicating the planet's place, and the elapsed days. Thus the reason for the rule is clear.

Similarly, to find the number of solar days which have elapsed from the beginning of a *kalpa* up to any given epoch, suppose S, y, denote respectively the *saura* days in the *kalpa* and the *saura* days elapsed, and A, a, the corresponding additive months. Now since one additive month occurs in every $32\frac{1}{2}$ solar months (*Goládhyáya*, IV, 9,10), we evidently get the proportion, $S : A :: y : a$, whence $a=\frac{Ay}{S}=x+\frac{b}{S}$ suppose, x being an integer, and $\frac{b}{S}$ a proper fraction. Consequently, $\frac{Ay-b}{S}=x$, and a positive integral solution of this equation will give the solar days and the integral number of additive months that have elapsed. Since by supposition, y is less than S, and x less than A, we can show as above that there can be only *one* positive integral solution, satisfying the conditions of the problem.

The case of finding the elapsed lunar days from the given remainder of deficient days or *avamas*, is precisely similar to the above, it being observed that an *avama* occurs in $64\frac{1}{11}$ lunar days (*Goládhyáya*, IV, 12).

In a period of $32\frac{1}{2}$ solar months there are $33\frac{1}{2}$ lunar months very nearly; this excess of the number of lunar months, *viz.*, one lunar month is called an *adhimása* or additive month, because a proportionate multiple of it is to be added to the solar months in any given period in order to convert them into lunar months. Again, in a period of $64\frac{1}{11}$ lunar days there are $63\frac{1}{11}$ terrestrial or mean solar days very nearly; this difference between the two numbers, *viz.*, one mean solar day, is called an *avama* or subtractive day, because a proportionate multiple of it is to be subtracted from the lunar days in any given period in order to convert them into mean solar days.]

265. Rule for a conjunct pulverizer.[1] If the divisor be the same and the multipliers various, then, making the sum of those multipliers the dividend, and the sum of the remainders a single remainder, and applying the foregoing method of investigation, the precise multiplier so found is denominated a conjunct one.

---

[1] *Sanslishta-kuttaka* or *sanslishtasphuta-kuttaka*, a distinct pulverizing multiplier belonging to conjunct residues.—Gan. A multiplier deduced from the sum of multipliers and that of remainders.—Súr.

[The reason for the rule will appear from the solution of the example which follows.]

266. Example. What quantity is it, which multiplied by five, and divided by sixty-three, gives a residue of seven; and the same multiplied by ten and divided by sixty-three, a remainder of fourteen? Declare the number[1].

Here the sum of the multipliers is made the dividend, and the sum of the residues, a subtractive quantity; and the statement is as follows:—

$\frac{\text{dividend } 15}{\text{divisor } 63}$. Subtractive 21. Or reduced to least terms:—

$\frac{\text{dividend } 5}{\text{divisor } 21}$. Subtractive 7.

Proceeding as before,[2] the multiplier is found 14.

[In this example we have two simultaneous equations involving three unknown quantities. Let y = quantity required. Then we have evidently, $5y=63m+7$, $10y=63n+14$, where m and n are certain positive integers. Put $m+n=x$; thus, $63x-15y=-21$, whence y can be found by §256, and the reason for the rule in §265 is obvious.]

---

[1] [See *Golúdhyáya*, XIII, 13—15.—Ed.]

[2] The quotient as it comes out in this operation is not to be taken: but it is to be separately sought with the several original multipliers applied to this quantity and divided by the divisor as given.—Gan.

CHAPTER XIII.

COMBINATION OF DIGITS.[1]

267. Rule.[2] The product of multiplication of the arithmetical series beginning and increasing by unity and continued to the number of places, will be the variations of number with specific figures: that divided by the number of digits and multiplied by the sum of the digits, being repeated in the places of figures and added together, will be the sum of the permutations.

[Let there be n digits. Then evidently there are $\underline{|n}$ numbers which can be formed with all these digits. Consider any one of these digits, and denote it by d. In $\underline{|n-1}$ cases, d is in the units' place in as many cases d is in the tens' place, in as many cases d is in the hundreds' place, and so on. Thus the sum arising from the d alone is $\underline{|n-1}\{d+10d+100d+......10^{n-1}d\}$. Proceeding similarly with the other digits, we get the sum of all the numbers $=\underline{|n-1}\times$ sum of digits $\times(10^{n-1}+...+10+1)=\frac{\underline{|n}}{n}\times$ sum of digits $\times(10^{n-1}+...+10+1)$, which stated in words leads to the rule. The meaning of the phrase, *being repeated in the places of figures and added together*, is obvious See foot-note to §13.]

[1] *Anka pása-vyavahára*, concatenation of digits: a mutual mixing of the numbers, as it were a rope of numbers, their variations being likened to a coil. See Gan. and Súr.

[2] To find the number of the permutations and the sum or amount of them, with specific numbers.—Gan. and Súr.

268. Example. How many variations of number can there be with two and eight, or with three, nine and eight, or with the continued series from two to nine? Tell promptly the several sums of these numbers.

Statement of the first example: 2, 8. Here the number of places is 2. The product of the series from 1 to the number of places and increasing by unity, will be 2. Thus the permutations of number are found 2. That product 2, multiplied by the sum of the figures 10, is 20; and divided by the number of digits 2, is 10. This repeated in the places of figures $(^{10}_{10})$ and added together, is 110, the sum of the numbers.

Statement of the second example: 3, 9, 8.

The arithmetical series is 1, 2, 3, of which the product is 6; and so many are the variations of number. That multiplied by the sum 20, is 120; which divided by the number of digits 3, gives 40; and this, repeated in the three places of figures and summed, makes 4440, the sum of the numbers.

Statement of the third example: 2, 3, 4, 5, 6, 7, 8, 9. The arithmetical series beginning and increasing by unity is 1, 2, 3, 4, 5, 6, 7, 8. The product gives the permutation of numbers, 40320. This, multiplied by the sum of the figures 44, is 1774080, which divided by the number of terms 8, is 221760; and the quotient being repeated in the eight places of figures and summed, the total is the sum of the numbers, 2463999975360.

269. Example. How many are the variations of form of the god *Sambhu* by the exchange of his ten attributes held reciprocally in his several hands, namely, the rope, the elephant's hook, the serpent, the tabor, the skull, the trident, the bedstead, the dagger, the arrow, and the bow[1] as those of *Hari* by the exchange of the mace, the discus, the lotus and the conch?

Statement: number of places 10.

In the same mode, as above shown, the variations of form are found 3628800. So the variations of form of *Hari* are 24.

270. Rule.[2] The permutations found as before,

---

[1] *Sambhu* or *Siva* is represented with ten arms, and holding in his ten hands the ten weapons or symbols here specified; and, by changing the several attributes from one hand to another, a variation may be effected in the representation of the idol, in the same manner as the image of *Hari* or *Vishnu* is varied by the exchange of his four symbols in his four hands. The twenty-four different representations of *Vishnu*, arising from this diversity in the manner of placing the weapons of attributes in his four hands, are distinguished by as many discriminative titles of the god allotted to those figures in the theogonies or *Purá nas*. It does not appear that distinct titles have been in like manner aassigned to any of the more than three millions of varied represonatations of *Siva*.

The ten attributes of *Siva* are :—1st, *pása*, a rope or chain for binding an elephant; 2nd, *ankusa*, a hook for guiding an elephant; 3rd, *ahi*, a serpent; 4th, *damaru*, a tabor: 5th, *kapála*, a human skull; 6th, *trisúla*, a trident; 7th, *khatwánga*, a bedstead, or a club in the form of the foot of one; [it may also mean a club having a skull at the top.—Ed.] 8th, *sakti*, a dagger; 9th, *sara* an arrow; 10th, *chápa*, a bow.

[2] Special; being applicable when two or more of the digits are alike.

being divided by the permutations separately computed for as many places as are filled by like digits, will be the variations of number, from which the sum of the numbers will be found as before.

[Let there be n digits ; and suppose p of them to be d_1, q of them to be d_2, and the rest unlike, namely, d_3, d_4, &c. Then the variations of number will clearly be $= \dfrac{\lfloor n}{\lfloor p \lfloor q}$.

(See Todhunter's Algebra, Art. 497.) The mumber of cases in which d_1 is in the units' or hundreds' &c. place is $\dfrac{\lfloor n-1}{\lfloor p-1 \lfloor q}$ (Todhunter's Algebra, Art. 497) ; and hence the sum arising from d_1 alone is $\dfrac{\lfloor n-1}{\lfloor p-1 \lfloor q}(10^{n-1}+\cdots+10+1)d_1$. Similarly the sum arising from d_2 is $\dfrac{\lfloor n-1}{\lfloor p \lfloor q-1}(10^{n-1}+\ldots\ldots+10+1)d_2$, and that arising from d_3, d_4, &c., is

$$\frac{\lfloor n-1}{\lfloor p \lfloor q}(10^{n-1}+\ldots\ldots+10+1)(d_3+d_4+\&c.)$$

Hence the sum of all the numbers is

$$\frac{\lfloor n}{n}(10^{n-1}+\ldots\ldots+10+1)\left\{\frac{d_1}{\lfloor p-1 \lfloor q}+\frac{d_2}{\lfloor p \lfloor q-1}+\frac{d_3+d_4+\&c.}{\lfloor p \lfloor q}\right\}$$

$$=\frac{\lfloor n}{n\lfloor p \lfloor q}\quad(pd_1+qd_2+d_3+d_4+\&c.)\quad(10^{n-1}+\ldots\ldots+10+1)$$

$$=\frac{\lfloor n}{n\lfloor p \lfloor q}\times\text{sum of digits}\times(10^{n-1}+\cdots+10+1),$$ whence the rule.]

271. Example. How many are the numbers with two, two, one and one ? And tell me quickly,

mathematician, their sum : also with four, eight, five, five and five, if thou be conversant with the rule of permutation of numbers.

Statement of the 1st example : 2, 2, 1, 1. Here the permutations found as before (§267) are 24. First, two places are filled by like digits (2, 2), and the permutations for that number of places are 2. Next two other places are filled by like digits (1, 1), and the permutations for these places are also 2. Total 4. The permutations 24 divided by 4 give 6 for the variations of number ; *viz.*,[1] 2211, 2121, 2112, 1212, 1221, 1122. The sum[2] of the numbers is found as before 9999.

Statement of the 2nd example : 4, 8, 5, 5, 5. Here the permutations found as before are 120, which, divided by the permutations for three place, *viz.*, 6, give the variations 20 : *viz.*, 48555, 84555, 54855, 58455, 55485, 55845, 55548, 55584, 45855, 45585, 45558, 85455, 85545, 85554, 54585, 58545, 55458, 55854, 54558, 58554. The sum[3] of the numbers comes out 1199988.

272. Rule[4] : half a stanza. The series of the

---

[1] The enumeration of the possible combinations is termed *prastára*.

[2] The variations 6, multiplied by the sum of the figures 6, and divided by the number of digits 4, give 9 ; which being repeated in four places of figures and summed, makes 9999.

[3] The variations 20, multiplied by the sum of the figures 27, give 540, which, divided by the number of digits 5, makes 108 : and this being repeated in five places of figures and summed, yields 1199988.

[4] To find the variations for a definite number of places with indeterminate digits.—Gan.

numbers decreasing by unity from the last[1] to the number of places, being multiplied together, will be the variations of number, with dissimilar digits.

[This rule gives the ordinary formula for the number of permutations of n things taken r at a time, *viz.*, $n(n-1)(n-2)......(n-r+1)$.]

273. Example. How many are the variations of number with any digits except cipher exchanged in six places of figures? If thou know, declare them.

The last number is nine. Decreasing by unity, for as many as are the places of figures, the statement of the series is 9. 8. 7. 6. 5. 4. The product of these is 60480.

274. Rule[2] : two stanzas. If the sum of the digits be determinate, the arithmetical series of numbers from one less than the sum of the digits, decreasing by unity, and continued to one less than the places, being divided by one and so forth, and the quotients being multiplied together, the product will be equal to the variations of number. This rule must be understood to hold good, provided the sum of the digits be less than the number of places added to nine.

A compendium only has been here delivered for fear of prolixity, since the ocean of calculation has no bounds.

---

[1] That is, from nine [in the example which follows.]—Gan.

[2] To find the permutations with indeterminate digits for a definite sum and a specific number of places.—Gan.

[Let s = the sum of the digits,

n = the number of digits,

and let $s = n + m$.

Then by supposition, $n+m<n+9$, or $m<9$, or $m+1$ not >9, so that even if $n-1$ of the n digits be 1's, the remainder of the sum, $m+1$ being not >9, can form the remaining digit.

Now let the n 1's composing n be denoted by 1^n, 1^{n-1}, 1^{n-2} ....1^1, and the m 1's composing m, by 1_1, 1_2, $1_3 \cdots \cdots 1_m$. Then, if we fix 1^n in the first place on the left, and take the different permutations of the remaining $n-1+m$ symbols 1^{n-1}, $1^{n-2}, \ldots\ldots 1^1$ and 1_1, 1_2, $1_3 \ldots\ldots 1_m$, of which the $n-1$ indexed 1's are considered to be alike and of one sort, and the other m 1's are considered to be alike and of another sort, and place each of these permutations to the right of 1^n, and regard the sum of each indexed 1 with the group of 1's with suffixes, if any, following it on its right as forming a digit of one of the required numbers,

we shall have a series of numbers like the following:—

$(1^n\, 1_1\, 1_2)\,(1^{n-1}\, 1_3\, 1_4\, 1_5) \ldots\ldots (1^p \ldots\ldots 1_{m-1}) \cdots (1^2\, 1_m)\,(1^1)$,

$(1^n\, 1_1\, 1_2\, 1_3\, 1_4)\,(1^{n-1}) \ldots\ldots (1^p \ldots\, 1_n) \ldots\ldots (1^2)\,(1^1)$,

....

This series will evidently contain all the required numbers and those alone; and the number of these numbers being the number required, the problem is reduced to finding the number of permutations of $n+m-1$ things taken all together, of which $n-1$ are alike and of one sort, and m are alike and of another sort

And this number $= \dfrac{\underline{|n+m-1}}{\underline{|n-1}\;\underline{|m}}$

$$= \frac{(n+m-1)\,(n+m-1-1)\cdots(n+m-1-\overline{\ldots-1})\,(m-1)\cdots 1}{\underline{|n-1}\;\underline{|m}}$$

$$=\frac{(n+m-1)(n+m-2)\cdots(n+m-1-\overline{n-2})}{|\underline{n-1}}$$

$$=\frac{(s-1)(s-2)\ldots\ldots(s-\overline{n-1})}{1.\ 2.\ldots\ldots\ldots\ldots(n-1)},$$

which proves the rule.]

[The following solution is given in Mahendra Nath Ray's Algebra, Part II, page 374 :—

Consider the product of the n equal factors $(x^1+x^2+x^3+\cdots+x^9+\cdots)(x^1+x^2+x^3+\cdots+x^9+\ldots)$ $\times(x^1+x^2+x^3+\cdots+x^9+\cdots)\cdots$

Suppose we are required to find the coefficient to x^s in the product We shall get *one* x^s in the product corresponding to every possible way of so selecting a term out of each of the n factors that their product may be x^s, that is, that the sum of the indiccs of the n terms selected may be s. Hence corrcsponding to each way of obtaining an x^s in the product, we shall get a number consisting of n digits, so that the sum of the digits may be s, since from the condition that $s<n+9$, no power of x higher than the 9th can be taken from any of the factors to form x^s. Hence the total number of numbers that can be so formed

=the total number of ways of obtaining x^s in the product

=the coefficient of x^s in the product

=the coefficicnt of x^s in $(x+x^2+x^3+\ldots+x^9+\ldots)^n$

=the coefficient of x in $x^n(1+x+x^2+\ldots)^n$

=the coeffieient of x^{s-n} in $(1-x)^{-n}$

$$=\frac{n(n+1)(n+2)\cdots(n+s-n-1)}{|\underline{s-n}}=\frac{|\underline{s-1}}{|\underline{n-1}\,|\underline{s-n}}$$

$$=\frac{(s-1)(s-2)\cdots\{s-(n-1)\}}{1.\ 2\ldots\ldots(n-1)}.\ ]$$

275. Example. How many various numbers are there, with digits standing in five places, the sum of which is thirteen? If thou know, declare them.

Here the sum of the digits less one is 12. The decreasing series from this to one less than the number of digits, divided by unity, &c. being exhibited, the statement is, $\frac{12}{1}.\frac{11}{2}.\frac{10}{3}.\frac{9}{4}$. The product of their multiplication $(\frac{11880}{24})$ is equal to the variations of the number[1], 495.

276. Though neither multiplier nor divisor be asked, nor square, nor cube, still presumptuous inexpert scholars in arithmetic will assuredly fail in (problems on) this combination of numbers.

277. Joy and happiness is indeed ever increasing in this world for those who have *Lílávatí* clasped to their throats[2], decorated as the members are with neat reduction of fractions, multiplication and involution, pure and perfect as are the solutions, and tasteful as is the speech which is exemplified.

---

[1] 91111, 52222, 13333, each five ways; 55111, 22333, each ten ways; 82111, 73111, 64111, 43222, 61222, each twenty ways; 72211, 53311, 44221, 44311, each thirty ways; 63211, 54211, 53221, 43321, each sixty ways. Thus the total is 495.

[2] By constant repetition of the text. This stanza, ambiguously expressed and bearing a double import, implies a simile: as a charming woman closely embraced, whose person is embellished by an assemblage of elegant qualities, who is pure and perfect in her conduct, and who utters agreeable discourse. See Gan.

[In Pandit Jívánanda Vidyáságara's edition of the original, there is a stanza after the above, showing the varied scholarship of Bháskara, and stating that the present work was composed by him. It was probably added by some pupil of Bháskara, and so it has been omitted by Colebrooke. It runs as follows:—"The author of this (*Líláwati*) is that illustrious Bháskara, (a scholar) of vast erudition, who thoroughly mastered eight works on grammar, (*viz.*, those of Indra, Chandra, Kásakritsni, Ápisali, Sákatáyana, Pánini, Amara, and Jainendra),[1] six works on medical science, (*viz.*, *Agnivesa-sanhitá*, *Bheda-sanhitá*, *Játúkarna-sanhitá*, *Parásara-sanhitá*, *Sirapáni-sanhitá* and *Hárita-sanhitá*),[2] the six philosophical systems (*viz.*, *Sánkhya*, *Yoga*, *Nyáya*, *Vaiseshika*, *Mímánsá* and *Vedánta*), five works on *ganita* (calculation), (*viz.*, *Paulisa-siddhánta*, *Romaka-siddhánta*, *Básishtha-siddhánta*, *Súrya-siddhánta* and *Paitámaha-siddhánta*),[3] and the four *Vedas* (*viz.*, the *Rik*, the *Yajush*, the *Sáman* and the *Atharvan*); and who understood the three *Ratnas*, (*i.e.*, the three *Prasthánas* of the *Vedánta*, *viz.*, the *Sútras*, the *Upanishads* and the *Prakaranas*), as well as the two *Mímánsás*,[4] and the one eternal *Brahman*, the aim and scope of both."]

[1 See Bibliotheca Indica, Nirukta, Vol. IV, Appendix, page *jau*. Some of these authors composed dictionaries and not works on grammar. Thus the word *vyákaranáni* in the original has been rather loosely used.—ED.]

[2 These six ancient works form the basis of the later works of Charaka, Susruta and Bágbhata. The works of Charaka and Susruta are usually called *sanhitás*; that of Bágbhata is known under the name of *Ashtángá-hridaya*.—ED.]

[3 See Varáha-mihira's *Vrihat-sanhitá*, Ch. II.—ED.]

[4 Namely, the *Púrva-mímánsá* of Jaimini, usually called the *Mímánsá*, and the *Uttara-mímánsá* of Vyása, usually called the *Vedánta*. —ED.]

APPENDIX.

लीलावती ।

—:o:—

[प्रथमोऽध्यायः ।]

प्रीतिं भक्तजनस्य यो जनयते विघ्नं विनिघ्नन् स्मृतस्तं वृन्दारकवृन्दवन्दितपदं नत्वा मतङ्गाननम्। पाटीं सद्गणितस्य वच्मि चतुरप्रीतिप्रदां प्रस्फुटां सङ्क्षिप्ताक्षरकोमलामलपदैर्लालित्यलीलावतीम् ॥ १ ॥

वराटकानां दशकद्वयं [२०] यत् सा काकिणी ताश्च पणश्चतस्रः। ते षोडश द्रम्म इहावगम्यो द्रम्मैस्तथा षोडशभिश्च निष्कः ॥ २ ॥

तुल्या यवाभ्यां कथितात्र गुञ्जा वल्लस्त्रिगुञ्जो धरणं च तेऽष्टौ। गद्याणकस्तद्द्वयमिन्द्रतुल्यैः [१४] वल्लैस्तथैको घटकः प्रदिष्टः ॥ ३ ॥

दशार्द्धगुञ्जं प्रवदन्ति माषं माषाह्वयैः षोडशभिश्च कर्षम्। कर्षैश्चतुर्भिश्च पलं तुलाज्ञाः कर्षं सुवर्णस्य सुवर्णसंज्ञम् ॥ ४ ॥

यवोदरैरङ्गुलमष्टसङ्ख्यैर्हस्तोऽङ्गुलैः षड्गुणितैश्चतुर्भिः। हस्तैश्चतुर्भिर्भवतीह दण्डः क्रोशः सहस्रद्वितयेन तेषाम् ॥ ५ ॥

स्याद्योजनं क्रोशचतुष्टयेन तथा कराणां दशकेन वंशः। निवर्तनं विंशतिवंशसङ्ख्यैः क्षेत्रं चतुर्भिश्च भुजैर्निबद्धम् ॥ ६ ॥

हस्तोन्मितैर्विस्तृतिदैर्घ्यपिण्डैर्यद्द्वादशास्रं घनहस्तसंज्ञम् । धान्यादिके यद्घनहस्तमानं शास्त्रोदिता मागधखारिका सा ॥ ७ ॥

द्रोणस्तु खार्याः खलु षोडशांशः स्यादाढको द्रोणचतुर्थभागः । प्रस्थश्चतुर्थांश इहाढकस्य प्रस्थाङ्घ्रिराद्यैः कुडवः प्रदिष्टः ॥ ८ ॥

[पादोनगद्यानकतुल्यटङ्कैर्द्विसप्ततुल्यैः कथितोऽत्र सेरः । मणाभिधानं खयुगै-(४०) श्च सेरैर्धान्यादितौल्येषु तुरुष्कसंज्ञा ॥ १ ॥]

शेषा कालादिपरिभाषा लोकतः प्रसिद्धा ज्ञेया ।

इति परिभाषा ।

[द्वितीयोऽध्यायः ।]

[प्रथमः परिच्छेदः ।]

लीलागललुलल्लोलकालव्यालविलासिने । गणेशाय नमो नील-कमलामलकान्तये ॥ ९ ॥

एकदशशतसहस्रायुतलक्षप्रयुतकोटयः क्रमशः । अर्बुदमब्जं खर्वं निखर्वमहापद्मशङ्कवस्तस्मात् । जलधिश्चान्त्यं मध्यं परार्द्धमिति दशगुणोत्तराः संज्ञाः । सङ्ख्यायाः स्थानानां व्यवहारार्थं कृताः पूर्वैः ॥ १०-११ ॥

[द्वितीयः परिच्छेदः ।]

अथ सङ्कलितव्यवकलितयोः करणसूत्रं वृत्तार्द्धम् । कार्य्यः क्रमादुत्क्रमतोऽथवाङ्कयोगो यथा स्थानकमन्तरं वा ॥ १२ ॥

अत्रोद्देशकः । अये बाले लीलावति मतिमति ब्रूहि सहितान् द्विपञ्चद्वात्रिंशत्त्रिनवतिशताऽष्टादश दश । शतोपेतानेतानयुतवियुतांश्चापि वद मे यदि व्यक्ते युक्तिव्यवकलनमार्गेऽसि कुशला ॥ १३ ॥

न्यासः । २ । ५ । ३२ । १९३ । १८ । १० । १०० ।

संयोजनाज्जातम् । ३६० ।

अयुता (१००००) च्छोधिते जातम् । ९६४० ।

इति सङ्कलितव्यवकलिते ।

गुणने करणसूत्रं सार्द्धवृत्तद्वयम् । गुण्यान्त्यमङ्कं गुणकेन हन्यादुत्सारितेनैवमुपान्तिमादीन् । गुण्यस्त्वधोऽधोगुणखण्डतुल्यस्तैः खण्डकैः सङ्गुणितो युतो वा । भक्तोगुणः शुध्यति येन तेन लब्ध्या च गुण्यो गुणितः फलं वा । द्विधा भवेद्रूपविभाग एवं स्थानैः पृथग्वा गुणितः समेतः । इष्टोनयुक्तेन गुणेन निघ्नोऽभीष्टघ्नगुण्यान्वितवर्जितो वा ॥

१४-१५ ॥

अथोद्देशकः । बाले बालकुरङ्गलोलनयने लीलावति प्रोच्यतां पञ्चत्र्येकमितादिवाकरगुणा अङ्काः कतिस्युर्यदि । रूपस्थानविभागखण्डगुणने कल्यासि कल्याणिनि छिन्नास्तेन गुणेन ते च गुणिता जाताः कति स्युर्वद ॥ १६ ॥

न्यासः । गुण्यः । १३५ । गुणकः । १२ ।

गुण्यान्त्यमङ्कं गुणकेन हन्यादिति कृते जातम् । १६२० ।

अथवा गुणरूपविभागे खण्डे कृते । ८ । ४ । आभ्यां पृथग्गुण्ये गुणिते युते च जातं तदेव । १६२० ।

अथवा गुणकस्त्रिभिर्भक्तोलब्धं । ४ । एभिस्त्रिभि-(३) श्च गुण्ये गुणिते जातं तदेव । १६२० ।

अथवा स्थानविभागे खण्डे । १ । २ । आभ्यां पृथग्गुण्ये गुणिते यथास्थानयुते च जातं तदेव । १६२० ।

अथवा इष्टोनेन गुणेन (१०) द्वाभ्याञ्च (२) पृथग्गुण्ये गुणिते युते च जातं तदेव । १६२० ।

अथवाष्टयुतेन गुणेन (२०) गुण्ये गुणितेऽष्ट-(८) गुणितगुण्यहीने च जातं तदेव । १६२० ।

इति गुणनप्रकारः ।

भागहारे करणसूत्रं वृत्तम्। भाज्याद्धरः शुध्यति यद्गुणः स्यादन्त्यात् फलं तत् खलु भागहारे। समेन केनाप्यपवर्त्त्य हारभाज्यौ भजेद्वा सति सम्भवे तु॥ १७॥

अत्र पूर्वोदाहरणे गुणिताङ्कानां स्वगुणच्छेदानां भागहारार्थं न्यासः भाज्यः। १६२०।
भाजकः। १२।

भजनाल्लब्धो गुणः। १३५।

अथवा भाज्यहारौ त्रिभिरपवर्त्तितौ $\frac{५४०}{४}$ चतुर्भिर्वा $\frac{४०५}{३}$ स्वस्व हारेण हृते फलं तदेव। १३५।

इति भागहारः।

वर्गे करणसूत्रं वृत्तद्वयम्। समद्विघातः कृतिरुच्यतेऽथ स्थाप्योऽन्त्यवर्गो द्विगुणान्त्यनिघ्नाः। स्वस्वोपरिष्टाच्च तथा परेऽङ्कास्त्यक्त्वान्त्यमुत्सार्य्य पुनश्च राशिम्॥ खण्डद्वयस्याभिहतिर्द्विनिघ्नी तत् खण्डवर्गैक्ययुता कृतिर्वा। इष्टोनयुग्राशिवधः कृतिः स्यादिष्टस्य वर्गेण समन्वितो वा॥ १८-१९॥

अत्रोद्देशकः। सखे नवानाञ्च चतुर्दशानां ब्रूहि त्रिहीनस्य शतत्रयस्य। पञ्चोत्तरस्याप्ययुतस्य वर्गं जानासि चेद्वर्गविधानमार्गम्॥ २०॥

न्यासः। ९। १४। २९७। १०००५।

एषां यथोक्तकरणेन जातावर्गाः। ८१। १९६। ८८२०९। १००१०००२५।

अथवा नवानां खण्डे। ४। ५। अनयोराहति—। २०। द्विनिघ्नी। ४०। तत्खण्डवर्गैक्येन। ४१। युता जाता सैव कृतिः। ८१।

अथवा चतुर्दशानां खण्डे। ६। ८। अनयोराहति—। ४८। द्विनिघ्नी। ९६। तत्खण्डवर्गौ। ३६। ६४। अनयोरैक्येन। १००। युता जाता सैव कृतिः। १९६।

अथवा खण्डे। ४। १०। तथापि सैव कृतिः। १९६।

अथवा राशिः। २९७। अयं त्रिभिरूनितः पृथग्युतश्च। २९४। ३००। अनयोर्घातः। ८८२००। त्रिवर्ग-। ९। युतो जातोवर्गः स एव। ८८२०९।

एवं सर्वत्रापि।

इति वर्गः।

वर्गमूले करणसूत्रं वृत्तम्। त्यक्त्वान्त्याद्विषमात् कृतिं द्विगुणयेन्मूलं समे तद्धृते त्यक्त्वा लब्धकृतिं तदाद्यविषमाल्लब्धं द्विनिघ्नं न्यसेत्।

पङ्क्त्यां पङ्क्तिहृते समेऽन्यविषमावाप्तावर्गं फलं पङ्क्त्यां तद्द्विगुणं न्यसेदिति मुहुः पङ्क्तेर्दलं स्यात्पदम् ॥ २१ ॥

अत्रोद्देशकः। मूलं चतुर्णाञ्च तथा नवानां पूर्वं शतानाञ्च सखे कृतीनाम्। पृथक् पृथग्वर्गपदानि विद्धि बुद्धेर्विवृद्धिर्यदि तेऽत्र जाता ॥ २२ ॥

न्यासः। ४। ९। ८१। १९६। ८८२०९। १००१०००२५।

लब्धानि क्रमेण मूलानि। २। ३। ९। १४। २९७। १०००५।

इति वर्गमूलम्।

घने करणसूत्रं वृत्तत्रयम्। समत्रिघातश्च घनः प्रदिष्टः स्थाप्यो-घनोऽन्त्यस्य ततोऽन्त्यवर्गः। आदित्रिनिघ्नस्तत आदिवर्गस्त्र्यन्त्याहतो-ऽथादिघनश्च सर्वे॥ स्थानान्तरत्वेन युता घनः स्यात् प्रकल्प्य तत्-खण्डयुगं ततोऽन्त्यम्। एवं मुहुर्वर्गघनप्रसिद्धा वाद्याङ्कतो वा विधिरेष कार्य्यः॥ खण्डाभ्यां वाहतोराशिस्त्रिघ्नः खण्डघनैक्ययुक्। वर्गमूल-घनस्वघ्नो वर्गराशिर्घनो भवेत् ॥ २३-२५ ॥

अत्रोद्देशकः। नवघनं त्रिघनस्य घनं तथा कथय पञ्चघनस्य घनश्च मे। घनपदञ्च ततोऽपि घनात् सखे यदि घनेऽस्ति घना भवतो मतिः ॥ २७ ॥

न्यासः । ९ । २७ । १२५ ।

जाताः क्रमेण घनाः । ७२९ । १९६८३ । १९५३१२५ ।

अथवा राशिः । ९ । अस्य खण्डे । ४ । ५ । आभ्यां राशिर्हतः । १८० । त्रिनिघ्नश्च । ५४० । खण्डघनैक्येन । १८९ । युतो जातो घनः । ७२९ ।

अथवा राशिः । २७ । अस्य खण्डे । २० । ७ । आभ्यां हतस्त्रिघ्नश्च । ११३४० । खण्डघनैक्येन । ८३४३ । युतो जातो घनः । १९६८३ ।

अथवा राशिः । ४ । अस्य मूलम् । २ । घनः । ८ । अयं स्वघ्नो-जातश्चतुर्णां घनः । ६४ ।

अथवा राशिः । ९ । अस्य मूलं । ३ । घनः । २७ । अस्य वर्गोनवानां घनः । ७२९ । य एव वर्गराशिघनः स एव वर्गमूलघनवर्गः ।

इति घनः ।

अथ घनमूले करणसूत्रं वृत्तद्वयम् । आद्यं घनस्थानमथाघने द्वे पुनस्तथान्त्याद्घनतोविशोध्यम् । घनं पृथक्स्थं पदमस्य कृत्या त्रिघ्न्या तदाद्यं विभजेत् फलन्तु ॥ १९ ॥ पङ्क्त्यां न्यसेत्तत्कृतिमन्त्यनिघ्नीं त्रिघ्नीं त्यजेत्तत्प्रथमात् फलस्य । घनं तदाद्याद्घनमूलमेवं पङ्क्ति-र्भवेदेव मतः पुनश्च ॥ २७-२८ ॥

अत्र पूर्वोद्देशके उक्तघनानां मूलार्थं न्यासः । ७२९ । १९६८३ । १९५३१२५ ।

क्रमेण लब्धानि मूलानि । ९ । २७ । १२५ ।

इति घनमूलम् ।

इति परिकर्म्माष्टकं समाप्तम्

[तृतीयः परिच्छेदः ।]

अथ भिन्नपरिकर्म्माष्टकम् । तत्रादावंशसवर्णनम् । तत्रापि भागजातौ करणसूत्रं वृत्तम् । अन्योऽन्यहाराभिहतौ हरांशौ राश्योः समच्छेदविधानमेवम् । मिथो हराभ्यामपवर्त्तिताभ्यां यद्वा हरांशौ सुधियात्र गुण्यौ ॥ २९ ॥

अत्रोद्देशकः । रूपत्रयं पञ्चलवस्त्रिभागो योगार्थमेतान्वद तुल्यहारान् । त्रिषष्टिभागश्च चतुर्दशांशः समच्छिदौ मित्र वियोजनार्थम् ॥ ३० ॥

न्यासः $\frac{३}{१} + \frac{१}{५} + \frac{१}{३}$

जाता समच्छेदाः $\frac{४५}{१५} + \frac{३}{१५} + \frac{५}{१५}$

योगे जातम् $\frac{५३}{१५}$

अथ द्वितीयोदाहरणार्थं न्यासः $\frac{१}{१४} - \frac{१}{६३}$

एतौ समापवर्त्तिताभ्यां हाराभ्यां। २। ८।

मिथस्संगुणितौ समच्छेदौ $\frac{९}{१२६}$ — $\frac{२}{१२६}$

वियोजिते जातम् $\frac{७}{१२६}$

समापवर्त्तिते च जातम् $\frac{१}{१८}$

इति भागजातिः।

अथ प्रभागजातौ करणसूत्रं वृत्तार्द्धम्। लवालवघ्नाश्च हराहरघ्ना भागप्रभागेषु सवर्णनं स्यात्॥ ३१॥

अत्रोद्देशकः। द्रम्मार्द्धत्रिलवद्वयस्य सुमते पादत्रयं यद्भवेत्तत्पञ्चांशकषोड़शांशचरणः सम्प्रार्थितेनार्थिने। दत्तो येन वराटकाः कति कदर्य्येणार्पितास्तेन मे ब्रूहि त्वं यदि वेत्सि गणिते जातिं प्रभागाभिधाम्॥ ३२॥

न्यासः $\frac{१}{१}$ $\frac{१}{२}$ $\frac{२}{३}$ $\frac{३}{४}$ $\frac{१}{५}$ $\frac{१}{१६}$ $\frac{१}{४}$

सवर्णिते जातम् $\frac{६}{७६८००}$

षड्भिरपवर्त्तिते जातम् $\frac{१}{१२८००}$

एको दत्तो वराटकः।

इति प्रभागजातिः।

अथ भागानुबन्धभागापवाहयोः करणसूत्रं सार्धवृत्तम्। छेदघ्न-
रूपेषु लवा धनर्णमेकस्य भागा अधिकोनकाश्चेत्। स्वांशाधिकोनः
खलु यत्र तत्र भागानुबन्धे च लवापवाहे। तलस्थहारेण हरं निहन्यात्
स्वांशाधिकोनेन तु तेन भागान् ॥ ३३ ॥

अत्रोद्देशकः। साङ्घ्रिद्वयं त्रयं व्यङ्घ्रि कीदृग् ब्रूहि सवर्णितम्।
जानास्यंशानुबन्धं चेत्तथा भागापवाहनम् ॥ ३४ ॥

न्यासः २ ३
१/४ १/४

सवर्णिते जातम् ९/४ ११/४

अत्रोद्देशकः। अङ्घ्रिः स्वत्र्यंशयुक्तः स निजदलयुतः कीदृशः
कीदृशौ द्वौ त्र्यंशौ स्वाष्टांशहीनौ तदनु च रहितौ स्वत्रिभिः सप्तभागैः।
अर्धं स्वाष्टांशहीनं नवभिरथ युतं सप्तमांशैः स्वकीयैः कीदृक् स्याद्ब्रूहि
वेत्सि त्वमिह यदि सखेंऽशानुबन्धापवाहौ ॥ ३५ ॥

न्यासः १/४ २/३ १/२
१/३ १˙/८ १˙/८
१/२ ३˙/७ ९/७

क्रमेण सवर्णिते जातम् १/२ १/३ १/१

इति जातिचतुष्टयम्।

अथ भिन्नसङ्कलितव्यवकलितयोः करणसूत्रं वृत्तार्धम्। योगोऽन्तरं
तुल्यहरांशकानां कल्प्यो हरो रूपमहारराशेः ॥ ३६ ॥

अत्रोद्देशकः। पञ्चांशपादत्रिलवार्धषष्ठानेकीकृतान् ब्रूहि सखे ममैतान्। एभिश्च भागैरथ वर्जितानां किं स्यात्त्रयाणां कथयाशु शेषम्॥ ३७॥

न्यासः $\frac{१}{५}$ $\frac{१}{४}$ $\frac{१}{३}$ $\frac{१}{२}$ $\frac{१}{६}$

ऐक्ये जातम् $\frac{२९}{२०}$

अथैतैर्विवर्जितानां त्रयानां शेषम् $\frac{३१}{२०}$

इति भिन्नसङ्कलितव्यवकलिते।

अथ भिन्नगुणने करणसूत्रं वृत्तार्धम्। अंशाहतिश्छेदवधेन भक्ता लब्धं विभिन्ने गुणने फलं स्यात्॥ ३८॥

अत्रोद्देशकः। सत्र्यंशरूपद्वितयेन निघ्नं ससप्तमांशद्वितयं भवेत् किम्। अर्धं त्रिभागेण हतञ्च विद्धि दक्षोऽसि भिन्ने गुणनाविधौ चेत्॥ ३८॥

न्यासः २ २

$\frac{१}{३}$ $\frac{१}{७}$

सवर्णिते जातम् $\frac{७}{३}$ $\frac{१५}{७}$

गुणिते जातम् $\frac{५}{१}$

न्यासः $\frac{१}{२}$ $\frac{१}{३}$

गुणिते जातम् $\frac{१}{६}$

इति भिन्नगुणनम्।

अथ भिन्नभागहारे करणसूत्रं वृत्तार्द्धम्।

छेदं लवञ्च परिवर्त्त्य हरस्य शेषः कार्य्योऽथ भागहरणे गुणनाविधिश्च ॥ ४० ॥

अत्रोद्देशकः। सत्र्यंशरूपद्वितयेन पञ्च त्र्यंशेन षष्ठं वद मे विभज्य। दर्भीयगर्भाग्रसुतीक्ष्णबुद्धिश्चेदस्ति ते भिन्नहृतौ समर्था ॥४१॥

न्यासः २

$\frac{१}{३}$ $\frac{५}{१}$ $\frac{१}{३}$ $\frac{२}{६}$

यथोक्तकरणेन जातम् $\frac{२५}{७}$ $\frac{७}{२}$

इति भिन्नभागहारः।

अथ भिन्नवर्गादौ करणसूत्रं वृत्तार्द्धम्। वर्गे कृतौ घनविधौ तु घनौ विधेयौ हारांशयोरथ पदे च पदप्रसिद्धै ॥ ४२ ॥

अत्रोद्देशकः। सार्द्धत्रयाणां कथयाशु वर्गं वर्गात्ततो वर्गपदञ्च मित्र। घनञ्च मूलञ्च घनात्ततोऽपि जानासि चेद्वर्गघनौ विभिन्नौ ॥ ४३ ॥

न्यास ३

$\frac{१}{२}$

छेदघ्नरूपे कृते जातम् $\frac{७}{२}$

अस्य वर्गः $\frac{४९}{४}$ मूलं $\frac{७}{२}$

घनः $\frac{३४३}{८}$ अस्य मूलं $\frac{७}{२}$

इति भिन्नपरिकर्म्माष्टकम्।

[चतुर्थः परिच्छेदः।]

अथ शून्यपरिकर्म्मसु करणसूत्रमार्य्याद्वयम्। योगे खं क्षेपसमं वर्गादौ खं खभाजितो राशिः। खहरः स्यात् खगुणः खं खगुणश्चिन्त्यश्च शेषविधौ॥ शून्ये गुणके जाते खं हारश्चेत् पुनस्तदा राशिः। अविकृत एव ज्ञेयस्तथैव खेनोनितश्च युतः॥ ४४—४५॥

अत्रोद्देशकः। खं पञ्चयुग्भवति किं वद खस्य वर्गं मूलं घनं घनपदं खगुणाश्च पञ्च। खेनोद्धृता दश च कः खगुणो निजार्द्धयुक्तः स्त्रिभिश्च गुणितः खहृतस्त्रिषष्टिः॥ ४६॥

न्यासः। ०। एतत्पञ्चयुतं जातम्। ५। खस्य वर्गः। ०। मूलम्। ०। घनम्। ०। तन्मूलम्। ०।

न्यासः। ५। एते खेन गुणिता जाताः। ०।

न्यासः। १०। एते खभक्ताः $\frac{१०}{०}$

अज्ञातो राशिस्तस्य गुणः । ० । सार्धं क्षेपः $\frac{1}{2}$ गुणः । ३ । हरः । ० ।
दृश्यं । ६३ । ततो वक्ष्यमाणेन विलोमविधिना इष्टकर्मणा वा लब्धो
राशिः । १४ । अस्य गणितस्य ग्रहगणिते महानुपयोगः ।

इति शून्यपरिकर्माष्टकम् ।

—*—

[तृतीयोऽध्यायः ।]

[प्रथमः परिच्छेदः ।]

अथ व्यस्तविधौ करणसूत्रं वृत्तद्वयम् । छेदं गुणं गुणं छेदं वर्गं मूलं पदं कृतिम् । ऋणं स्वं स्वमृणं कुर्याद् दृश्ये राशिप्रसिद्धये ॥ अथ स्वांशाधिकोने तु लवाढ्योनोहरोहरः । अंशस्त्वविकृतस्तत्र विलोमे शेषमुक्तवत् ॥ ४७—४८ ॥

अत्रोद्देशकः । यस्त्रिघ्नस्त्रिभिरन्वितः स्वचरणैर्भक्तास्ततः सप्तभिः स्वत्र्यंशेन विवर्जितः स्वगुणितो हीनो द्विपञ्चाशता । तन्मूलेऽष्टयुते हृतेऽपि दशभिर्जातं द्वयं ब्रूहि तं राशिं वेत्सि हि चञ्चलाक्षि विमलां बाले विलोमक्रियाम ॥ ४९ ॥

न्यासः । गुणः । ३ । क्षेपः $\frac{3}{4}$ भाजकः । ७ । ऋणं $\frac{1}{3}$

वर्गः ।—। ऋणम् । ५२ । मूलम् ।—। क्षेपः । ८ । हरः । १० ।

दृश्यम् । २ ।

यथोक्तकरणेन जातो राशिः ॥ २८ ॥

इति व्यस्तविधिः ।

[द्वितीयः परिच्छेदः ।]

अथेष्टकर्मसु करणसूत्रं वृत्तम् । उद्देशकालापवदिष्टराशिः क्षुण्णो हृतोऽशैः रहितो युतो वा । इष्टाहतं दृष्टमनेन भक्तं राशि-र्भवेत् प्रोक्तमितीष्टकर्म ॥ ५० ॥

अत्रोद्देशकः । पञ्चघ्नः स्वत्रिभागोनो दशभक्तः समन्वितः । राशित्र्यंशार्द्धपादैः स्यात् को राशिर्द्व्यूनसप्ततिः ॥ ५२ ॥

न्यासः । गुणः । ५ । स्वत्रिभागः $\frac{१}{३}$ हरः । १० । राश्यंशाः $\frac{१}{३}$ $\frac{१}{२}$ $\frac{१}{४}$ दृश्यम् । ६८ ।

अत्र किल कल्पितराशिः । ३ । पञ्चघ्नः । १५ । स्वत्रिभागोनः । दशभक्तः । १ । कल्पित- । ३ । राशेस्त्र्यंशार्द्धपादैः $\frac{३}{३}$ $\frac{३}{२}$ $\frac{३}{४}$ समन्वितो हरो जातः $\frac{१७}{४}$ अथ दृष्टम् । ६८ । इष्टेन । ३ । गुणितम् । १०४ । हरेण $\frac{१७}{४}$ भक्तं जातो राशिः । ४८ ।

एवं सर्वत्रोदाहरणे राशिः केनचिद्गुणितो भक्तो वा राश्यंशेन रहितो

युतो वा दृष्टस्तत्रेष्टं राशिं प्रकल्प्य तस्मिन्नुद्देशकालापवत्कर्म्मणि कृते यन्निष्पद्यते तेन भजेद्दृष्टमिष्टगुणं फलं राशिः स्यात्।

[अत्र दृष्टजात्युदाहरणम्। यूथार्द्धं सत्रिभागं वनविचरहतं कुञ्जराणाञ्च दृष्टं षड्भागश्चैव नद्यां पिवति च सलिलं सप्तमांशेन मिश्रः। पद्मिन्या चाष्टमांशः स्वनवमसहितः क्रीड़ते सानुरागो नागेन्द्रो हस्तिनीभिस्त्रिभिरनुगतः का भवेद्यूथसङ्ख्या ॥ १ ॥

न्यासः १/२ १/६ १/८ दृश्यम्। ४।

१/३ १/७ १/९

एषां सवर्णनं द्वाभ्यामपवर्त्तितं २/३ ४/२१ ५/३६

पुनरेषां सवर्णितानामैक्यं नवभिरपवर्त्तितं २५१/२५२

इष्टोनं १/२५२

अनेन दृष्टे। ४। इष्ट गुणिते भक्ते जाता हस्तिसङ्ख्या। १००८।]

अपरोदाहरणम्। अमलकमलराशेस्त्र्यंशपञ्चांशषष्ठैस्त्रिनयनहरिसूर्य्या येन तुर्य्येण चार्य्या गुरुपदमथ षड्भिः पूजितं शेषपद्मैः सकलकमलसङ्ख्यां क्षिप्रमाख्याहि तस्य ॥ ५२ ॥

न्यासः १/३ १/५ १/६ १/४ दृश्यम्। ६।

अत्रेष्टं राशिं। १। प्रकल्प्याप्राग्वज्जातोराशिः। १२०।

अन्यदुदाहरणम्। हारस्तारस्तरुण्या निधुवनकलहे मौक्तिकानां विशीर्णो भूमौ यातस्त्रिभागः शयनतलगतः पञ्चमांशोऽस्य दृष्टः। प्राप्तः षष्ठः सुकेश्या गणक दशमकः संगृहीतः प्रियेण दृष्टं षट्कञ्च सूत्रे कथय कतिपयैर्मौक्तिकैरेष हारः ॥ २ ॥

न्यास १/३ १/५ १/६ १/१० दृश्यम्। ६।

अत्रेष्टं राशिं। १। प्रकल्प्यप्राग्वज्जातोराशिः। ३०।

अथ शेषजात्युदाहरणम्। स्वार्धं प्रादात्प्रयागे नवलवयुगलं योऽवशेषाच्च काश्यांशेषाङ्घ्रिं शुल्कहेतोः पथि दशमलवान् षट् च शेषाद्गयायाम्। शिष्टा निष्कत्रिषष्टिर्निजगृहमनया तीर्थपान्थः प्रयातस्तस्य द्रव्यप्रमाणं वद यदि भवता शेषजातिः श्रुतास्ति ॥ ५२ ॥

न्यासः १/१
•१/२ दृश्यम्। ६३।
•२/९
•१/४
६/१०

अत्र रूपम्। १। राशिं प्रकल्प्य भागान् शेषान् शेषादपास्य अथवा भागापवाहविधिना सवर्णिते जातम् ७/६०

अनेन दृष्टे। ६३। दृष्टगुणिते भक्ते जातं द्रव्यप्रमाणम्। ५४०।

इदं विलोमसूत्रेणापि सिध्यति।

अथ विश्लेषजात्युदाहरणम्। पञ्चांशोऽलिकुलात्कदम्बमगमत् त्र्यंशः शिलीन्ध्रं तयोर्विश्लेषस्त्रिगुणो मृगाक्षि कुटजं दोलायमानोऽपरः। कान्ते केतकमालतीपरिमलप्राप्तैककालप्रियादूताहूत इतस्ततो भ्रमति खे भृङ्गोऽलिसङ्ख्यां वद ॥ ५४ ॥

न्यासः १/५ १/३ २/५ दृश्यम्। १।

जातमलिकुलमानम्। १५। एवमन्यत्रापि।

इतीष्टकर्म्म।

[तृतीयः परिच्छेदः।]

सङ्क्रमणे करणसूत्रं वृत्तार्द्धम्। योगोऽन्तरेणोनयुतोऽर्द्धितस्तौ राशी स्मृतौ सङ्क्रमणाख्यमेतत् ॥ ५५ ॥

अत्रोद्देशकः। ययोर्योगः शतं सैकं वियोगः पञ्चविंशतिः। तौ राशी वद मे वत्से वेत्सि सङ्क्रमणं यदि ॥ ५६ ॥

न्यासः। योगः। १०१। अन्तरम्। २५। जातौ राशी। ३८। ६३।

वर्गसङ्क्रमणे करणसूत्रं वृत्तार्द्धम्। वर्गान्तरं राशिवियोगभक्तं योगस्ततः प्रोक्तवदेव राशिः ॥ ५७ ॥

उद्देशकः। राश्योर्ययोर्वियोगोऽष्टौ तत्कृत्योश्च चतुःशती। विवरं वद तौ राशी शीघ्रं गणितकोविद ॥ ५८ ॥

न्यासः। राश्यन्तरम्। ८। कृत्यन्तरम्। ४००।

जातौ राशी। २१। २९।

इति विषमकर्म।

[चतुर्थः परिच्छेदः]

अथ किञ्चिद्वर्गकर्म प्रोच्यते। इष्टकृतिरष्टगुणिताव्येकादलिता· विभाजितेष्टेन। एकः खादस्य कृतिर्दलिता सैका परो राशिः॥ रूपं द्विगुणेष्टहृतं सेष्टं प्रथमोऽथवा परो रूपम्। कृतियुतिवियुतीव्येके वर्गौ स्यातां ययो राशयोः॥ ५९—६०॥

उद्देशकः। राशयोर्ययोः कृतिवियोगयुतीनिरेके मूलप्रदे प्रवद तौ मम मित्र यत्र। क्लिश्यन्ति वीजगणिते पटवोऽपि मूढाः षोढोक्तगूढगणितं परिभाषयन्तः॥

अत्र प्रथमानयने,कल्पितमिष्टम् $\frac{१}{२}$

अस्य कृतिः $\frac{१}{४}$

अष्टगुणा जाता। २। इयं व्येका। १।

दलिता $\frac{१}{२}$

इष्टेन $\frac{१}{२}$ हृतो जातः प्रथमो राशिः। १। अस्य कृतिः। १।

दलिता $\frac{१}{२}$

सेका $\frac{३}{२}$ अयमपरो राशिः।

एवमेतौ राशी $\frac{१}{१}$ $\frac{३}{२}$

एवमेकेनेष्टेन जातौ राशी $\frac{७}{२}$ $\frac{५७}{८}$

द्विकेन $\frac{३१}{४}$ $\frac{९६३}{३२}$

अथ द्वितीयप्रकारेणेष्टम्। १। अनेन द्विगुणेन। २। रूपम्। १।

भक्तं $\frac{१}{२}$

इष्टेन सहितं जातः प्रथमो राशिः ३/२

द्वितीयो रूपम्। १।

एवं राशी ३/२ १/१

एवं द्विकेन ६/४ १/३

त्रिकेन १८/६ १/१

संगुणेन ११/८ १/१

अथवा सूत्रम्। इष्टस्य वर्गवर्गौ घनश्च तावष्टसंगुणौ। प्रथमः सेकोराशी स्यातां व्यक्तोऽथवाव्यक्तौ ॥ ६१ ॥

इष्टं १/२ अस्य वर्गवर्गः १/१६

अष्टघ्नः १/२

सेको जातः प्रथमो राशिः ३/२

पुनरिष्टं १/२ अस्य घनः १/८

अष्टगुणो जातो द्वितीयो राशिः १/१

एवं जातौ राशी ३/२ १/१

अथैकेष्टेन। ९। ८।

द्विकेन। १२९। ६४।

त्रिकेण। ६४९। २२६।

एवं सर्वेष्वपि इष्टवशादानन्त्यम्। पाटीसूत्रोपमं वीजं गूढमित्यवभासते। नास्ति गूढममूढानां नैव षोढेत्यनेकधा॥ अस्ति त्रैराशिकं

पाटीबीजस्य विमला मतिः। किमज्ञातं सुबुद्धीनामतो मन्दार्थमुच्यते॥

इति वर्गकर्म।

[पञ्चमः परिच्छेदः।]

अथ गुणकर्म। तत्र दृष्टमूलजातौ करणसूत्रं वृत्तद्वयम्। गुणघ्नमूलोनयुतस्य राशेर्दृष्टस्य युक्तस्य गुणार्द्धकृत्या। मूलं गुणार्द्धेन युतं विहीनं वर्गीकृतं प्रष्टुरभीष्टराशिः। यदा लवैश्चोनयुतस्य राशिरेकेन भागोनयुतेन भक्ता। दृश्यंत्र तथा मूलगुणश्च ताभ्यां साध्यस्ततः प्रोक्तवदेव राशिः॥ ६२—६३॥

यो राशिः मूलेन केनचिद्गुणितेन ऊनो दृष्टस्तस्य गुणार्द्धकृत्या युक्तस्य दृष्टस्य यत्पदं तद्गुणार्द्धेन युक्तं कार्य्यं यदि गुणघ्नमूलयुतो दृष्टस्तर्हि हीनं कार्य्यं तस्य वर्गोराशिः स्यात्।

मूलोने दृष्टे तावदुदाहरणम्। बाले मरालकुलमूलदलानि सप्त तीरे विलासभरमन्थरगाण्यपश्यम्। कुर्वच्च केलिकलहं कलहंस-युग्मं शेषं जले वद मरालकुलप्रमाणम्॥ ६४॥

न्यासः। मूलगुणः $\frac{७}{२}$ दृश्यम्। २।

दृष्टस्यास्य। २। गुणार्द्धकृत्या $\frac{४९}{१६}$

युक्तस्य $\frac{८१}{१६}$

मूलम् ९/४

गुणार्द्धेन ७/४

युतम्। ४।

वर्गीकृतं जातं हंसकुलमानम्। १६।

अथ मूलयुते दृष्टे उदाहरणम्। स्वपदैर्नवभिर्युक्तः स्याच्चत्वारिंशताधिकम्। शतद्वादशकं विद्वन् कः स राशिर्निगद्यताम्॥ ६५॥

न्यासः। मूलगुणः। ९। दृश्यम्। १२४०।

गुणार्द्धस्य ९/२

कृत्या ८१/४

युतं जातम् ५०४१/४

अस्य मूलं ७१/२

गुणार्द्धेन ९/२

अत्र विहीनम्। ३१।

वर्गीकृतं जातो राशिः। ९६१।

भागमूलोने दृष्टे उदाहरणम्। जातं हंसकुलस्य मूलदशकं मेघागमे मानसं प्रोड्डीय स्थलपद्मिनीवनमगादष्टांशकोऽभ्रस्तटात्।

वाले वालमृणालशालिनि जले केलिक्रियालालसं दृष्टं हंसयुगत्रयञ्च सकलां यूथस्य सङ्ख्यां वद ॥ ६६ ॥

न्यासः । मूलगुणः । १० । भागः $\frac{१}{८}$ दृश्यम् । ६ । यदा लवैश्चोनयुत इत्युक्तत्वादत्रैकेन भागोनेन $\frac{७}{८}$ दृश्यमूलगुणौ भक्ता जातं दृश्यं $\frac{४८}{७}$

मूलगुणः $\frac{८०}{७}$

गुणार्द्धम् $\frac{४०}{७}$

अस्य कृत्या $\frac{१६००}{४९}$

युक्तं दृश्यम् $\frac{१६३६}{४९}$

अस्य मूलं $\frac{४४}{७}$

गुणार्द्धेन $\frac{४१}{७}$

युतं वर्गीकृतं जातो हंसराशिः । १४४ ।

उदाहरणम् । पार्थः कर्णवधाय मार्गणगणं क्रुद्धो रणे सन्दधे तस्यार्द्धेन निवार्य्य तच्छरगणं मूलैश्चतुर्भिर्हयान् । शल्यं षड्भि-रथेषुभिस्त्रिभिरपि च्छत्रं ध्वजं कार्मुकं चिच्छेदास्य शिरः शरेण कति ते यान् अर्जुनः सन्दधे ॥ ६७ ॥

न्यासः । मूलगुणः । ४ । भागः $\frac{१}{२}$ दृश्यम् । १० ।

यदा लवैश्चोनयुत इत्यादिना जातं बाणमानम् ॥ १०० ॥

अपिच । अलिकुलदलमूलं मालतीं यातमष्टौ निखिलनवम-

भागास्त्वालिनो भृङ्गमेकम्। निशि परिमललुब्धं पद्ममध्ये निरुद्धं प्रतिरणति रणन्तं ब्रूहि कान्तेऽलिसङ्ख्याम्॥ ६८॥

अत्र किल राशिनवांशाकं राश्यर्द्धमूलस्य राशेरूर्णं रूपद्वयं दृश्यम्। एतदृणं दृश्यस्याङ्कितं राश्यर्द्धस्य भवतीति।

तथा न्यासः। मूलगुणकः $\frac{\dot{1}}{2}$ भागः $\frac{\dot{8}}{9}$ दृश्यम्। १।

प्राग्वल्लब्धं राशिदलम्। ३६। एतद्द्विगुणितमलिकुल-मानम्। ७२।

भागमूलयुते दृष्टे उदाहरणम्। यो राशिरष्टादशभिः स्वमूलैः राशित्रिभागेण समन्वितश्च। जातं शतद्वादशकं तमाशु जानीहि पाट्यां पटुतास्ति ते चेत्॥ ६९॥

न्यासः। मूलगुणकः। १८। भागः $\frac{1}{3}$ दृश्यम्। १२००।

अनेकेन भागयुतेन $\frac{4}{3}$

मूलगुणं दृश्यञ्च भक्त्वा प्राग्वज्जातो राशिः। ५७६।

इति गुणकर्म

[षष्ठः परिच्छेदः।]

अथ त्रैराशिके करणसूत्रं वृत्तम्। प्रमाणमिच्छा च समान-जाती आद्यन्तयोस्तः फलमन्यजाति। मध्ये तदिच्छाहतमाद्य-हृत्स्यादिच्छाफलं व्यस्तविधिर्विलोमे॥ ७०॥

उदाहरणम्। कुङ्कुमस्य सदलं पलद्वयं निष्कसप्तमलवैर्विलि-

भिर्यदि। प्राप्यते सपदि मे वणिग्वर ब्रूहि निष्कनवकेन तत्कियत् ॥ ७१ ॥

न्यासः $\frac{३}{७}$ $\frac{५}{२}$ $\frac{९}{१}$

लब्धानि कुङ्कुमपलानि। ५२। कर्षौ। २।

अपि च। प्रकृष्टकर्पूरपलत्रिषष्ट्या चेल्लभ्यते निष्कचतुष्कयुक्तम्। शतं तदा द्वादशभिः सपादैः पलैः किमाचक्ष्व सखे विचिन्त्य ॥ ७२ ॥

न्यासः $\frac{६३}{१}$ $\frac{१०४}{१}$ $\frac{४९}{४}$

इच्छागुणितं फलम् $\frac{५०९६}{४}$

छेदभक्तम्। १२७४। आद्येन। ६३। हृतं लब्धा निष्काः। २०। शेषम्। १४। षोडशगुणितम्। २२४। आद्येन भक्तं जाताद्रम्माः। ३। एवं पणाः। ८। काकिण्यः। ३। वराटकाः। भागाश्च $\frac{१}{९}$

पुनश्च। द्रम्मद्वयेन साष्टांशा शालितण्डुलखारिका लभ्या चेत्पण-सप्तत्या तत् किं सपदि कथ्यताम् ॥ ७३ ॥

अत्र प्रमाणेच्छयोः सजातीयकरणार्थं द्रम्मद्वयस्य पणीकृतस्य न्यासः $\frac{३२}{१}$ $\frac{९}{८}$ $\frac{७०}{१}$

लब्धे खार्यौ। २। द्रोणाः। ७। आढकः। १। प्रस्थौ। २।

इति त्रैराशिकम्।

अथ व्यस्तं त्रैराशिकम्। इच्छावृद्धौ फले ह्रासो ह्रासे वृद्धिश्च जायते। व्यस्तं त्रैराशिकं तत्र ज्ञेयं गणितकोविदैः॥ ७४॥

यत्र इच्छावृद्धौ फले ह्रासः ह्रासे वा फले वृद्धिस्तत्र व्यस्तं त्रैराशिकं स्यात्। तद्यथा—

जीवानां वयसो मौल्ये तौल्ये वर्णस्य हेमनि। भागहारे च राशीनां व्यस्तं त्रैराशिकं भवेत्॥ ७५॥

जीववयोमूल्ये उदाहरणम्। प्राप्नोति चेत् षोडशवत्सरा स्त्री द्वात्रिंशतं विंशतिवत्सरा किम्। द्विधूर्वहो निष्कचतुष्कमुक्षा प्राप्नोति धूः षट्कवहस्तदा किम्॥ ७६॥

न्यासः। १६। ३२। २०। लब्धम् २५ ३/५

द्वितीयन्यासः। २। ४। ६। लब्धम् १ १/३

वर्णीयसुवर्णतौल्ये उदाहरणम्। दशवर्णं सुवर्णं चेद्गद्यानकमवाप्यते। निष्केण तिथिवर्णन्तु तदा वद कियन्मितम्॥ ७७॥

न्यासः। १०। १। १५। लब्धम् २/३

राशिभागहरके उदाहरणम्। सप्ताढ़केन मानेन राशौ शस्यस्य मापिते। यदि मानशतं जातं तदा पञ्चाढ़केन किम्॥ ७८॥

न्यासः। ७। १००। ५। लब्धम्। १४०।

इति व्यस्तं त्रैराशिकम्।

अथ पञ्चराशिकादौ करणसूत्रं वृत्तम्। पञ्चसप्तनवराशिकादिकेऽन्योन्यपक्षनयनं फलच्छिदाम्। संविधाय बहुराशिजे बधे स्वल्पराशिबधभाजिते फलम्॥ ७९॥

उदाहरणम्। मासे शतस्य यदि पञ्चकलान्तरं स्याद्वर्षे गते भवति किं वद षोड़शानाम्। कालं तथा कथय मूलकलान्तराभ्यां मूलं धनं गणक कालफले विदित्वा॥ ८०॥

न्यासः १ १२
१०० १६
५ ०

अन्योऽन्यपक्षनयने कृते १ १२
१०० १६
० ५

बहूनां राशीनां वधः। ९६०।

अल्पराशिवधः। १००। अनेन भक्ते लब्धम्। ९।

शेषम् $\frac{६०}{१००}$

विंशत्यापवर्त्य $\frac{३}{५}$

जातं कलान्तरम् ९ $\frac{३}{५}$

छेदघ्नरूपेष्विति कृते जातम् $\frac{४८}{५}$

अथ कालज्ञानार्थं न्यासः

१	०
१००	१६
५	$\frac{४८}{५}$

अन्योन्यपक्षनयने कृते

१	०
१००	१६
४८	५
	५

बहूनां राशीनां वधः। ४८००।

अल्पराशिवधः। ४००।

अनेन भक्ते लब्धा मासाः। १२।

मूलधनार्थं न्यासः　१　　१२

१००　　०

५　　४८/५

पूर्ववल्लब्धं मूलधनम्। १६। एवं सर्वत्र।

उदाहरणम्। सत्र्यंशमासेन शतस्य चेत् स्यात् कलान्तरं पञ्च सपञ्चमांशाः। मासैस्त्रिभिः पञ्चलवाधिकैस्तत्सार्द्धद्विषष्टेः फलमुच्यतां किम्॥ ८१॥

न्यासः　१　　३

१/३　　१/५

१००　　६२

५　　१/२

१/५

छेदघ्न रूपेष्विति कृते　४/३　　१६/५

१००　　१२५/२

२६/५　　०

अन्योऽन्यपक्षनयने कृते　४　　१६

५　　३

१००　　१२५

२/५　　२६

अत्र बहुराशिवधे। १५६०००।

अल्पराशिवधेनानेन। २००००।

भक्ते लब्धं कलान्तरम् ७
४
५

छेदवरूपेष्विति कृते जातम् ३९/५

कालादिज्ञानार्थं पूर्ववत्।

अथ सप्तराशिकम्। विस्तारे त्रिकराः कराष्टकमिता दैर्घ्ये विचित्राश्च चेद्रूपैत्कटपट्टसूत्रपटिका अष्टौ लभन्ते शतम्। दैर्घ्ये सार्द्धकरत्रयापरपटी हस्तार्द्धविस्तारिणी तादृक किं लभते द्रुतं वद वणिग्वाणिज्यकं वेत्सि चेत्॥ ८२॥

न्यासः ३ १/२

८ ७/२

८ १

१०० ०

लब्धो निष्कः। ०। द्रम्माः। १४। पणाः। ८। काकिणी। १। वराटकाः। ६। वराटकत्रिभागौ २/३

अथ नवराशिकम्। पिण्डे येऽर्कमिताङ्गुलाः किल चतुर्वर्गाङ्गुला-विस्तृतौ पट्टादीर्घतया चतुर्दशकरास्त्रिंशल्लभन्ते शतम्। एता विस्तृतिपिण्डदैर्घ्यमितयो येषां चतुर्वर्जिताः पट्टास्ते वद मे चतुर्दश सखे मूल्यं लभन्ते कियत॥ ८३॥

न्यासः

१२	८
१६	१२
१४	१०
३०	१४
१००	०

लब्धं मूल्यं निष्काः १६ $\frac{२}{३}$

अथैकादशराशिकम्। पट्टा ये प्रथमोदितप्रमितयोगव्यूतिमात्रे मतास्तेषामानयनाय चेच्छकटिनां द्रम्माष्टकं भाटकः। अन्ये ये तदनन्तरं निगदिता माने चतुर्वर्जितास्तेषां का भवतीति भाटकमितिर्गव्यूतिषट्के वद ॥ ८४ ॥

न्यासः

१२	८
१६	१२
१४	१०
३०	१४
१	६
८	०

लब्धाः भाटकद्रम्माः। ८।

अथ भाण्डप्रतिभाण्डके करणसूत्रं वृत्तार्द्धम्। तथैव भाण्डप्रतिभाण्डके विधिर्विपर्य्ययस्तत्र सदा हि मूल्ये ॥ ८५ ॥

उदाहरणम्। द्रम्मेण लभ्यत इहाम्रशतत्रयं चेत् त्रिंशत्पणेन विपणौ वरदाडिमानि। आम्रैर्दशभिः कति दाडिमानि लभ्यानि तद्विनिमयेन भवन्ति मित्र॥ ८६॥

न्यासः १६ १
३०० ३०
१० ०

लब्धानि दाडिमानि। १६।

इति लीलावत्यां प्रकीर्णकानि।

[चतुर्थोऽध्यायः।]

[प्रथमः परिच्छेदः।]

अथ मिश्रकव्यवहारे करणसूत्रं सार्द्धवृत्तम्। प्रमाणकालेन हतं प्रमाणं विमिश्रकालेन हतं फलञ्च॥ स्वयोगभक्ते च पृथक्स्थिते ते मिश्राहते मूलकलान्तरे स्तः। यद्वेष्टकर्माख्यविधेस्तु मूलं मिश्राच्च्युतं तच्च कलान्तरं स्यात्॥ ८७-८८॥
१

उद्देशकः। पञ्चकेन शतेनाब्दे मूलं स्वं सकलान्तरम्। सहस्रं चेत्पृथक तत्र वद मूलकलान्तरे॥ ८९॥

न्यासः १ १२
१०० १०००
५ ०

लब्धे क्रमेण मूलकलान्तरे। ६२५। ३७५।

अथवेष्टकर्म्मणा कल्पितमिष्टं रूपम्। १। उद्देशकालापवदिष्ट-राशिरित्यादिकरणेन रूपस्य वर्षे कलान्तरम् ३/५

एतद्युतेन रूपेण ८/५

हृष्टे। १००। रूपगुणे भक्ते लब्धं मूलधनम्। ६२५। एतन्मिश्रात्। १०००। च्युतं कलान्तरम्। ३७५।

मिश्रान्तरे करणसूत्रं वृत्तम्। अथ प्रमाणैर्गुणिताः स्वकाला व्यतीतकालघ्नफलोद्धृतास्ते। स्वयोगभक्ताश्च विमिश्रनिघ्नाः प्रयुक्त-खण्डानि पृथग् भवन्ति ॥ ८० ॥

उद्देशकः। यत्पञ्चकत्रिकचतुष्कशतेन दत्तं खण्डैस्त्रिभिर्गणक निष्कशतं षडूनम्। मासेषु सप्तदशपञ्चसु तुल्यमाप्तं खण्डत्रयेऽपि हि फलं वद खण्डसङ्ख्याम् ॥ ८१ ॥

न्यासः १। ७। १। १०। १। ५।
१०० १०० १००
५ ३ ४

मिश्रधनम्। ९४।

लब्धानि यथाक्रमेण खण्डानि। २४। २८। ४२।

पञ्चराशिवत्करणेन समकालान्तरम् ८
२
५

अथ मिश्रान्तरे करणसूत्रं वृत्तार्द्धम्। प्रक्षेपकामिश्रहता विभक्ताः प्रक्षेपयोगेन पृथक् फलानि॥

अत्रोद्देशकः। पञ्चाशदेकसहिता गणकाष्टषष्टिः पञ्चोनिता नवतिरादिधनानि येषाम्। प्राप्ता विमिश्रितधनैस्त्रिशती त्रिभिस्तैर्वाणिज्यतो वद विभज्य धनानि तेषाम्॥ ८३॥

न्यासः। प्रक्षेपकाः। ५१। ६७। ८५। मिश्रधनम्। ३००।

जातानि धनानि। ७५। १००। १२५।

एतान्यादिधनोनानि लाभाः। २४। ३२। ४०।

अथवा मिश्रधनं। ३००। आदिधनैक्येन। २०४। ऊनं सर्वलाभयोगः। ९६। अस्मिन् प्रक्षेपगुणिते प्रक्षेपयोग। १०४। भक्ते लाभाः। २४। ३२। ४०।

[द्वितीयः परिच्छेदः।]

वाप्यादिपूरणे करणसूत्रं वृत्तार्द्धम्। भजेच्छिदोंऽशैरथ तैर्विमिश्रैरूपं भजेत्स्यात्परिपूर्त्तिकालः॥ ८४॥

उदाहरणम्। ये निर्झरा दिनदिनार्द्धतृतीयषष्ठैः संपूरयन्ति हि

पृथक् पृथगेव मुक्ताः। वापीं यदा युगपदेव सखे विमुक्तास्ते केन वासरलवेन तदा वदाशु॥ ९५॥

न्यासः $\frac{१}{१}$ $\frac{१}{२}$ $\frac{१}{३}$ $\frac{१}{४}$

लब्धो वापीपूरणकालो दिनांशः $\frac{१२}{२५}$

[तृतीयः परिच्छेदः।]

अथ क्रयविक्रये करणसूत्रं वृत्तम्। पण्यैः स्वमूल्यानि भजेत् स्वभागैर्हत्वा तदैक्येन भजेच्च तानि। भागांश्च मिश्रेण धनेन हत्वा मूल्यानि पण्यानि यथाक्रमं स्युः॥ ९६॥

उद्देशकः। सार्धं तण्डुलमानकत्रयमहो द्रम्मेण मानाष्टकं मुद्गानाञ्च यदि त्रयोदशमिता एता वणिक्काकिणी। आदायार्पय तण्डुलांशयुगलं मुद्गैकभागान्वितं क्षिप्रं क्षिप्रभुजो व्रजेम हि यतः सार्थोऽग्रतो यास्यति॥ ९७॥

न्यासः। पण्ये $\frac{७}{२}$ $\frac{८}{१}$

स्वमूल्ये $\frac{१}{१}$ $\frac{१}{१}$

स्वभागौ $\frac{२}{१}$ $\frac{१}{१}$

मिश्रधनम् $\frac{१३}{१६}$

अत्र स्वमूल्ये स्वभागगुणिते पण्याभ्यां भक्ते जाते $\frac{४}{७}$ $\frac{१}{८}$ अनयोर्योगेन $\frac{३९}{५६}$ एते $\frac{४}{७}$ $\frac{१}{८}$ भागौ च $\frac{२}{१}$ $\frac{१}{१}$ मिश्रधनेन $\frac{१३}{१६}$ सङ्गुण्य भक्ते जाते तण्डुलमुद्गमूल्ये $\frac{२}{३}$ $\frac{७}{४८}$

तथा तण्डुलमुद्गमाने भागौ $\frac{१}{१२}$ $\frac{१}{२४}$

अत्र तण्डुलमूल्ये पणौ। २। काकिण्यौ। २। वराटकाः। १३। भागश्च $\frac{१}{३}$

मुद्गमूल्ये काकिण्यौ। २। वराटकाः। ६। वराटकत्रिभागौ $\frac{२}{३}$

उदाहरणम्। कर्पूरस्य वरस्य निष्कयुगलेनैकं पलं प्राप्यते वैश्यानन्दनचन्दनस्य च पलं द्रम्माष्टभागेन चेत्। अष्टांशेन तथाऽगुरोः पलदलं निष्केण मे देहि तान् भागैरेककषोड़शाष्टकमितैर्धूपं चिकीर्षाम्यहम्॥ ९८॥

न्यासः। पण्यानि $\frac{१}{१}$ $\frac{१}{१}$ $\frac{१}{२}$

मूल्यानि $\frac{३२}{१}$ $\frac{१}{८}$ $\frac{१}{८}$

भागाः $\frac{१}{१}$ $\frac{१६}{१}$ $\frac{८}{१}$

मिश्रधनं द्रम्माः। १६।

लब्धानि कर्पूरादीनां मूल्यानि १४ $\frac{२}{९}$ ० $\frac{८}{९}$ ० $\frac{८}{९}$

तथैव तेषां पण्यानि ० $\frac{४}{९}$ ७ $\frac{१}{९}$ ३ $\frac{५}{९}$

[चतुर्थः परिच्छेदः]

रत्नमिश्रे करणसूत्रं वृत्तम्। नरघ्नदानोनितरत्नशेषैरिष्टे हृते स्युः खलु मूल्यसङ्ख्याः। शेषैर्हृते शेषवधे पृथक्स्थैरभिन्नमूल्यान्यथवा भवन्ति ॥ ९९ ॥

अत्रोद्देशकः। माणिक्याष्टकमिन्द्रनीलदशकं मुक्ताफलानां शतं सद्वज्राणि च पञ्च रत्नवणिजां येषां चतुर्णां धनम्। सङ्गस्नेहवशेन ते निजधनाद्दत्त्वैकमेकं मिथो जातास्तुल्यधनाः पृथग्वद सखे तद्रत्नमूल्यानि मे ॥ १०० ॥

न्यासः। मा। ८। नी। १०। मु। १००। व। ५। दानम्। १। नराः। ४।

नरगुणितदानेन। ४। रत्नसङ्ख्यासूनितासु शेषाणि मा। ४। नी। ६। मु। ९६। व। १। एतैरिष्टराशौ भक्ते रत्नमूल्यानि स्युरिति। तानि च यथा कथञ्चिदिष्टे कल्पिते भिन्नानि भतोऽत्रेष्टं तथा सुधिया कल्प्यते यथाऽभिन्नानीति तथात्रापीष्टं कल्पितम्। ९६। अतो जातानि मूल्यानि। २४। १६। १। ९६। समधनम्। २२१।

अथवा शेषाणां घाते। २३०४। पृथक् शेषैर्भक्ते जातान्य-

भिन्नानि। ५७६। ३८४। २४। २३०४। जनानां चतुर्णां तुल्य-धनम्। ५५९२। तेषामिति द्रम्मासम्भाव्यन्ते ॥

[पञ्चमः परिच्छेदः।]

अथ सुवर्णगणिते करणसूत्रं वृत्तम्। सुवर्णवर्णाहतियोगराशौ स्वर्णैक्यभक्ते कनकैक्यवर्णः। वर्णो भवेच्छोधितहेमभक्ते वर्णाहृते शोधितहेमसङ्ख्या ॥ १०१ ॥

उदाहरणानि। विश्वार्करुद्रदशवर्णसुवर्णमाषाः दिग्वेदलोचनयुग-प्रमिताः क्रमेण। आवर्त्तितेषु वद तेषु सुवर्णवर्णस्तूर्णं सुवर्णगणितज्ञ वणिग् भवेत् कः ॥ १०२ ॥

ते शोधने यदि च विंशतिरत्नमाषाः स्युः षोड़श द्रविणवर्णमिति-स्तदा का। तच्छोधितं भवति षोड़शवर्णहेम ते विंशतिः कति भवन्ति तदा तु माषाः ॥ १०२ ॥

न्यासः
१३	१२	११	१०
१०	४	२	४

जाता आवर्त्तिते सुवर्णवर्णमितिः। १२।

एत एव यदि शोधिताः सन्तः षोड़श माषाः भवन्ति तदा वर्णः। १५।

यदि तदेव शोधितं षोड़शवर्णं स्वर्णं भवति तदा पञ्चदश माषाः भवन्ति।

अथ वर्णज्ञानाय करणसूत्रं वृत्तम्। स्वर्णैक्यनिघ्नाद्युतिजातवर्णात्सुवर्णतद्वर्णवधैक्यहीनात्। अज्ञातवर्णाग्निजसङ्ख्ययाप्तमज्ञातवर्णस्य भवेत् प्रमाणम्॥ १०४॥

उदाहरणम्। दशेशवर्णा वसुनेत्रमाषा अज्ञातवर्णस्य षडेतदैक्ये। जातं सखे द्वादशकं सुवर्णमज्ञातवर्णस्य वद प्रमाणम्॥ १०५॥

न्यासः
$\begin{matrix} १० & ११ & ० \\ ८ & २ & ६ \end{matrix}$

लब्धमज्ञातवर्णमानम्। १५।

सुवर्णज्ञानाय करणसूत्रं वृत्तम्। स्वर्णैक्यनिघ्नो युतिजातवर्णः स्वर्णघ्नवर्णैक्यवियोजितश्च। अहेमवर्णाग्निजयोगवर्णविश्लेषभक्तो विदिताग्निजं स्यात्॥ १०६॥

उदाहरणम्। दशेन्द्रवर्णागुणचन्द्रमाषाः किञ्चित्तथा षोडशकस्य तेषाम्। जातं युतौ द्वादशकं सुवर्णं कतीह ते षोडशवर्णमाषाः॥ १०७॥

न्यासः
$\begin{matrix} १० & १४ & १६ \\ ३ & १ & ० \end{matrix}$

लब्धं माषमानम्। १।

सुवर्णज्ञानायान्यत्करणसूत्रं वृत्तम्। साध्येनोनोऽनल्पवर्णो विधेयः साध्यो वर्णः स्वल्पवर्णोनितश्च। इष्टक्षुण्णे शेषके स्वर्णमाने स्यातां स्वल्पानल्पयोर्वर्णयोस्ते॥ १०८॥

उदाहरणम्। हाटकगुटिके षोडशदशवर्णे तद्युती सखे जातम्।
द्वादशवर्णसुवर्णं ब्रूहि तयोः स्वर्णमाने मे ॥ १०९ ॥

न्यासः $\frac{१६}{०}$ $\frac{१०}{०}$

साध्यो वर्णः। १२ ।

कल्पितमिष्टम्। १ ।

लब्धे सुवर्णमाने $\frac{१६}{२}$ $\frac{१०}{४}$

अथवा द्विकेनेष्टेन $\frac{१६}{४}$ $\frac{१०}{८}$

अर्द्धेनेष्टेन वा $\frac{१६}{१}$ $\frac{१०}{२}$

एवं बहुधा।

[षष्ठः परिच्छेदः।]

अथ छन्दश्चित्यादौ करणसूत्रं श्लोकत्रयम्। एकाद्येकोत्तरा अङ्का व्यस्ता भाज्याः क्रमस्थितैः। परः पूर्वेण सङ्गुण्यस्तत्परस्तत्परेण च ॥ एकद्वित्र्यादिभेदाः स्युरिदं साधारणं स्मृतम्। छन्दश्चित्युत्तरे छन्दस्युपयोगोऽस्य तद्विदाम् ॥ मूषावहनभेदादौ खण्डमेरौ च शिल्पके। वैद्यके रसभेदीये तन्नोक्तं विस्तृतेर्भयात् ॥ ११०—११२ ॥

तत्र छन्दश्चित्युत्तरे किञ्चिदुदाहरणम्। प्रस्तारे मित्र गायत्र्याः स्युः पादे व्यक्तयः कति। एकादि गुरवश्चाशु कथ्यतां तत्पृथक् पृथक्॥ ११३ ॥

इह हि षड़क्षरो गायत्रीचरणः अतः षड़न्तानां एकाद्येकोत्तराङ्कानां व्यस्तानां क्रमस्थानाञ्च

न्यासः
६ ५ ४ ३ २ १
१ २ ३ ४ ५ ६

यथोक्तकरणेन लब्धाः एकगुरुव्यक्तयः। ६। द्विगुरवः। १५। त्रिगुरवः। २०। चतुर्गुरवः। १५। पञ्चगुरवः। ६। षड्गुरवः। १। तथैका सर्वलघुः। १। एवमासामैक्यं पादव्यक्तिमितिः। ६४।

एवञ्चतुश्चरणाक्षरसङ्ख्यकानङ्कान् यथोक्तं विन्यस्य एकादिगुरुभेदानानीयतान् सेकानेकीकृत्य जाता गायत्रीवृत्तव्यक्तिसङ्ख्याः। १६७७७२१६।

एवमुक्थाद्युत्कृतिपर्य्यन्तं छन्दसां व्यक्तिमितिर्ज्ञातव्या।

उदाहरणं शिल्पे। एकद्वित्र्यादिमूषावहनमिति मह्यो ब्रूहि मे भूमिभर्त्तुर्हर्म्ये रम्येऽष्टमूषे चतुरविरचिते श्लक्ष्णशालाविशाले। एकद्वित्र्यादियुक्त्या मधुरकटुकषायाम्लकक्षारतिक्तैरेकस्मिन् षड्रसैः स्युर्गणक इति वद व्यञ्जने व्यक्तिभेदाः॥ ११४॥

न्यासः

८	७	६	५	४	३	२	१
१	२	३	४	५	६	७	८

लब्धा एकद्वित्र्यादिमूषावहनसङ्ख्याः

८	२८	५६	७०	५६	२८	८	१
१	२	३	४	५	६	७	८

एवमष्टमूषे राजगृहे मूषावहनभेदाः। २५५।

अथ द्वितीयोदाहरणे

न्यासः

६	५	४	३	२	१
१	२	३	४	५	६

लब्धा एकादिरससंयोगेन

पृथग व्यञ्जायः

६	२५	२०	१५	६	१
१	२	३	४	५	६

एतासामैक्यम्। ६३।

इति मिश्रकव्यवहारः समाप्तः।

[पञ्चमोऽध्यायः।]

—०—

अथ श्रेढीव्यवहारः।

[प्रथमः परिच्छेदः।]

तत्र सङ्कलितैक्ये करणसूत्रं वृत्तम्। सैकपदघ्नपदार्द्धमथैकाद्यङ्कयुतिः किल सङ्कलिताख्या। सा द्वियुतेन पदेन विनिघ्नी स्यात्त्रिहृता खलु सङ्कलितैक्यम्॥ ११५॥

उदाहरणम्। एकादीनां नवान्तानां पृथक् सङ्कलितानि मे। तेषां सङ्कलितैक्यानि प्रचक्ष्व तद्गुणक म्॥ ११६॥

न्यासः	१	२	३	४	५	६	७	८	९
सङ्कलितानि	१	३	६	१०	१५	२१	२८	३६	४५
एषामैक्यानि	१	४	१०	२०	३५	५६	८४	१२०	१६५

कृत्यादियोगे करणसूत्रं वृत्तम्। द्विघ्नपदं कुयुतं त्रिविभक्तं सङ्कलितेन हतं कृतियोगः। सङ्कलितस्य कृतेः सममेकाद्यङ्कघनैक्यमुदीरितमाद्यैः॥ ११७॥

उदाहरणम्। तेषामेव च वर्गैक्यं घनैक्यञ्च वद द्रुतम्। कृतिसङ्कलनामार्गे नाकुला यदि ते मतिः॥ ११८॥

न्यासः	१	२	३	४	५	६	७	८	९
वर्गैक्यम्	१	५	१४	३०	५५	९१	१४०	२०४	२८५
घनैक्यम्	१	९	३६	१००	२२५	४४१	७८४	१२९६	२०२५

यथोत्तरचयेऽन्त्यादिधनज्ञानाय करणसूत्रं वृत्तम्। व्येकपदघ्न-चयो मुखयुक् स्यादन्त्यधनं मुखयुग्दलितं तत्। मध्यधनं पदसङ्गु-णितं तत्सर्वधनं गणितञ्च तदुक्तम्॥ ११९॥

उदाहरणम्। आद्ये दिने द्रम्मचतुष्टयं यो दत्त्वा द्विजेभ्योऽ-नुदिनं प्रवृत्तः। दातुं सखे पञ्चचयेन पक्षे द्रम्मा वद द्राक्कति तेन दत्ताः॥ १२०॥

न्यासः। आ। ४। च। ५। ग। १५।

मध्यधनम्। ३९।

अन्त्यधनम्। ७४।

सर्वधनम्। ५८५।

उदाहरणान्तरम्। आदिः सप्त चयः पञ्च गच्छोऽष्टौ यत्र तत्र मे। मध्यान्त्यधनसङ्ख्ये के वद सर्वधनञ्च किम्॥ १२१॥

न्यासः। आ। ७। च। ५। ग। ८।

मध्यधनम् $\frac{४९}{२}$

अन्त्यधनम्। ४२।

सर्वधनम् । १८६ ।

समदिने गच्छे मध्यदिनाभावान्मध्यगाग्रागपरदिनधनयोर्योगार्धं मध्यदिनधनं भवितुमर्हतीति प्रतीतिरुत्पाद्या ।

मुखज्ञानाय करणसूत्रं वृत्तार्धम् । गच्छहृते गणिते वदनं स्याद्व्येकपदघ्नचयार्धविहीने ॥ १२२ ॥

उदाहरणम् । पञ्चाधिकं शतं श्रेढीफलं सप्त पदं किल । चयं त्रयं वयं विद्मो वदनं वद नन्दन ॥ १२३ ॥

न्यासः । आ । ० । च । ३ । ग । ७ । ध । १०५ ।
लब्धमादिधनम् । ६ ।

चयज्ञानाय करणसूत्रं वृत्तार्धम् । गच्छहृतं धनमादिविहीनं व्येकपदार्धहृतश्च चयः स्यात् ।

उदाहरणम् । प्रथममगमदह्ना योजने, योजनेशस्तदनु ननु कयासौ वृद्धि यातोऽध्वहृद्या । अरिकरिहरणार्थं योजनानामशीत्या रिपुनगरमवाप्तः सप्तरात्रेण धीमन् ॥ १२४ ॥

न्यासः । आ । २ । च । ० । ग । ७ । ध । ८० ।
लब्धमुत्तरम् $\frac{२२}{२१}$

गच्छज्ञानाय करणसूत्रं वृत्तम् । श्रेढीफलादुत्तरलोचनघ्नाच्चयार्द्ध-

वक्त्रान्तरवर्गयुक्तात्। मूलं मुखोनं चयखण्डयुक्तं चयोद्धृतं गच्छमुदाहरन्ति॥ १२५॥

उदाहरणम्। द्रम्मत्रयं यः प्रथमेऽह्नि दत्त्वा दातुं प्रवृत्तो द्विचयेन तेन। शतत्रयं षष्ट्यधिकं द्विजेभ्यो दत्तं कियद्भिर्दिवसैर्वदाशु॥ १२६॥

न्यासः। आ। ३। च। २। ग। ०। ध। ३६०। लब्धो गच्छः। १८।

[द्वितीयः परिच्छेदः।]

अथ द्विगुणोत्तरादिफलानयने करणसूत्रं सार्द्धवृत्तम्। विषमे गच्छे व्येके गुणकः स्थाप्यः समेऽर्द्धिते वर्गः। गच्छक्षयान्तमन्त्याद्व्यस्तं गुणवर्गजं फलं यत्तत्॥ व्येकं व्येकगुणोद्धृतमादिगुणं स्याद्गुणोत्तरे गणितम्॥ १२७॥

उदाहरणम्। पूर्वं वराटकयुगं येन द्विगुणोत्तरं प्रतिज्ञातम्। प्रत्यहमर्थिजनाय स मासे निष्कान् ददाति कति॥ १२८॥

न्यासः। आदिः। २। चयो गुणः। २। गच्छः। ३०।

लब्धावराटकाः। २१४७४८३६४६। निष्कवराटिकाभिर्भक्ताजातानिष्काः। १०४८५७। द्रम्माः। ९। पणाः। ९१ काकिण्यौ। २। वराटकाः। ६।

उदाहरणम्। आदिर्द्विकं सखे वृद्धिः प्रत्यहं त्रिगुणोत्तरा। गच्छः सप्तदिनं यत्र गणितं तत्र किं वद॥ १२९॥

न्यासः। आदिः। २। चयो गुणः। ३। गच्छः। ७।

लब्धं गणितम्। २१८६।

समादिवृत्तज्ञानाय करणसूत्रं सार्द्धार्य्या। पादाक्षरमितगच्छे गुणवर्गफलञ्चये द्विगुणे। समवृत्तानां सङ्ख्या तद्वर्गो वर्गवर्गश्च। स्वस्वपदोनौ स्यातामर्द्धसमानाञ्च विषमाणाम्॥ १३०—१३१॥

उदाहरणम्। समानामर्द्धतुल्यानां विषमाणां पृथक् पृथक्। वृत्तानां वद मे सङ्ख्यामनुष्टुप्छन्दसि द्रुतम्॥ १३२॥

न्यासः। उत्तरो गुणः। २। गच्छः। ८।

लब्धाः समवृत्तानां सङ्ख्याः। २५६।

तथार्द्धसमानाम्। ६५२८०।

विषमाणाञ्च। ४२९४९०१७६०।

इति श्रेढीव्यवहारः समाप्तः।

[षष्ठोऽध्यायः।]

अथ क्षेत्रव्यवहारः।

तत्र भुजकोटिकर्णानामन्यतमाभ्यामन्यतमानयनाय करणसूत्रं वृत्तद्वयम्। इष्टाद्बाहोर्यः स्यात्तत्स्पर्द्धिन्यां दिशीतरो बाहुः। त्र्यस्रे चतुरस्रे वा सा कोटिः कीर्त्तिता तज्ज्ञैः॥ १३३॥

तत्कृत्योर्योगपदं कर्णो दोःकर्णवर्गयोर्विवरात्। मूलं कोटिः कोटिश्रुतिकृत्योरन्तरात्पदं बाहुः॥ १३४॥

प्रकारान्तरेण तज्ज्ञानाय करणसूत्रं सार्द्धवृत्तम्। राश्योरन्तरवर्गेण द्विघ्ने घाते युते तयोः। वर्गयोगो भवेदेवं तयोर्योगान्तराहतिः। वर्गान्तरं भवेदेवं ज्ञेयं सर्वत्र धीमता॥ १३५॥

उदाहरणम्। कोटिश्चतुष्टयं यत्र दोस्त्रयं तत्र का श्रुतिः। कोटिं दोःकर्णतः कोटिश्रुतिभ्याञ्च भुजं वद॥ १३६॥

न्यासः

४

३

कोटिः। ४। भुजः। ३। भुजवर्गः। ९। कोटिवर्गः। १६। एतयोर्योगात्। २५। मूलम। ५। कर्णो जातः।

अथ कर्णभुजाभ्यां कोट्यनयनम्।

न्यासः

कर्णः। ५। भुजः। ३।
अनयोर्वर्गान्तरम्। १६।
एतन्मूलं कोटिः। ४।

अथ कोटिकर्णाभ्यां भुजानयनम्।

न्यासः

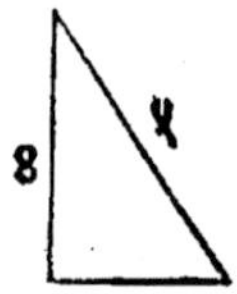

कोटिः। ४। कर्णः। ५।
अनयोर्वर्गान्तरम्। ९।
एतन्मूलं भुजः। ३।

कोटिश्चतुष्टयमिति पूर्वोक्तोदाहरणे।

न्यासः

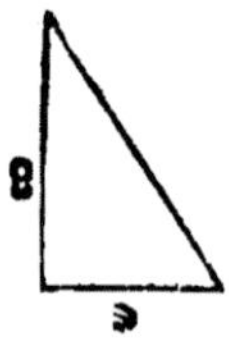

कोटिः। ४। भुजः। ३। अनयोर्घाते। १२। द्विघ्ने। २४। अन्तरवर्गेण। १। युते वर्गयोगः। २५। अस्य मूलं कर्णः। ५।

अथ कर्णभुजाभ्यां कोट्यानयनम्।

न्यासः

कर्णः। ५। भुजः। ३। अनयोर्योगः। ८। पुनरेतयोरन्तरेण। २। हतोवर्गान्तरम्। १६। अस्य मूलं कोटिः। ४।

अथ भुजज्ञानम्।

न्यासः

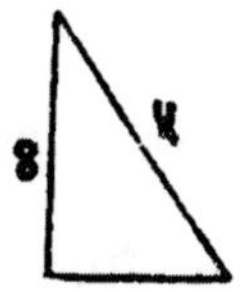

कोटिः। ४। कर्णः। ५। एवं जातो भुजः। ३।

उदाहरणम्। साङ्घ्रित्रयमितो बाहुर्यत्र कोटिश्च तादृशी। तत्र कर्णप्रमाणं किं गणक ब्रूहि मे द्रुतम्॥ १३७॥

न्यासः

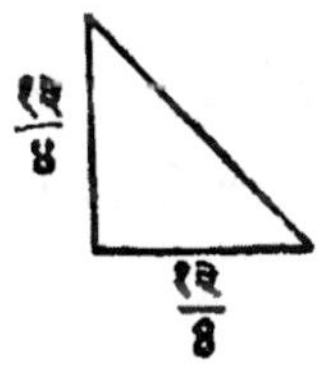

भुजः १३/४ कोटिः १३/४ अनयोर्वर्गयोर्योगः १६९/८ अस्य मूलाभावात् करणीगत एवायं कर्णः अस्यासन्नमूलज्ञानार्थमुपायः। वर्गेण महतेष्टेन हताच्छेदांशयोर्वधात्। पदं गुणपदक्षुण्णच्छिद्भक्तं निकटं भवेत्॥ १३८॥

अयं कर्णवर्गः १६९/८ अस्य छेदांशघातः। १३५२। अयुतघ्नः। १३५२००००। अस्यासन्नमूलम्। ३६७७। इदं गुणमूल। १००। गुणितच्छेदेन। ८००। भक्तं लब्धमासन्नपदम्। ४। भागाः ४७७/८०० अयं कर्णः। एवं सर्वत्र।

त्र्यस्रजात्ये करणसूत्रं वृत्तद्वयम्। इष्टो भुजोऽस्माद्द्विगुणेष्ट-

निघ्नादिष्टस्य कृत्यैकवियुक्तयाप्तम्। कोटिः पृथक् सेष्टगुणा भुजोना कर्णो भवेत् त्र्यस्रमिदन्तु जात्यम्॥ १३९॥

इष्टो भुजस्तत्कृतिरिष्टभक्ता द्विः स्थापितेष्टोनयुतार्द्धिता वा। तौ कोटिकर्णाविति कोटितो वा बाहुश्रुती वाऽकरणीगते स्तः॥ १४०॥

उदाहरणम्। भुजे द्वादशके यौ यौ कोटिकर्णावनेकधा। प्रकाराभ्यां वद क्षिप्रं तौ तावकरणीगतौ॥ १४१॥

न्यासः

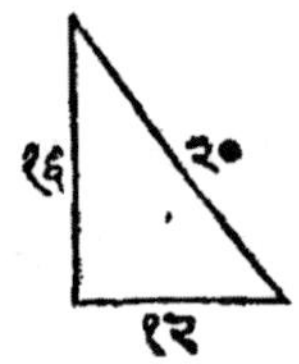

इष्टो भुजः। १२। इष्टम्। २। अनेन द्विगुणेन। ४। गुणितो भुजः। ४८। इष्ट। २। कृत्या। ४। एकोनया। ३। भक्तो लब्धा कोटिः। १६। इयमिष्टगुणा। ३२। भुजोना जातः कर्णः। २०।

त्रिकेनेष्टेन वा।

न्यासः

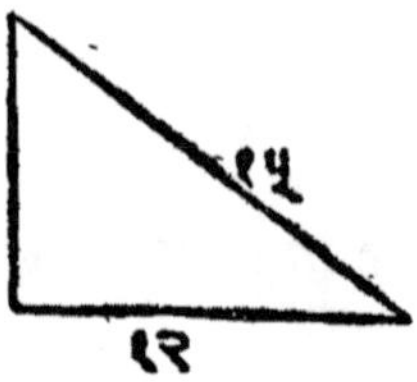

कोटिः। ९। कर्णः। १५।

पञ्चकेन वा ।

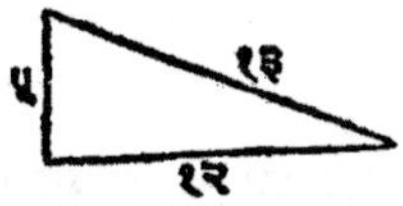

कोटिः । ५ । कर्णः । १३

इत्यादि ।

अथ द्वितीयप्रकारेण ।

न्यासः

इष्टो भुजः । १२ । अस्य कृतिः । १४४ । इष्टेन । २ । भक्ता लब्धम् । ७२ । इष्टेन । २ । ऊन । ७० । युता- । ७४ । वर्द्धितौ जातौ कोटिकर्णौ ३५ । ३७ ।

चतुष्टयेन वा ।

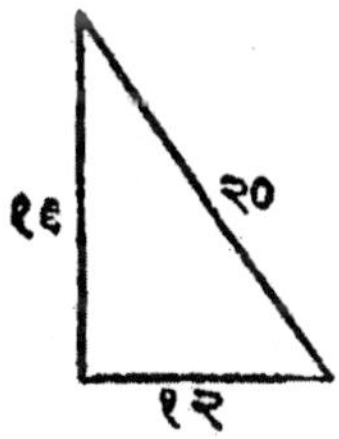

कोटिः । १६ ।

कर्णः । २० ।

षट्केन वा ।

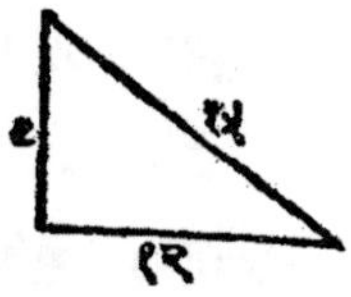

कोटिः । ९ ।

कर्णः । १५ ।

अथेष्टकर्णात् कोटिभुजानयने करणसूत्रं वृत्तम् । इष्टेन निघ्नाद्-

द्विगुणाच्च कर्णादिष्टस्य कृत्यैकयुजा यदाप्तम्। कोटिर्भवेत् सा पृथगिष्टनिघ्नी तत्कर्णयोरन्तरमत्र बाहुः ॥ १५२ ॥

उदाहरणम्। पञ्चाशीतिमिते कर्णे यौ यावकरणीगतौ। स्यातां कोटि भुजौ तौ तौ वद कोविद सत्वरम् ॥ १४३ ॥

न्यासः

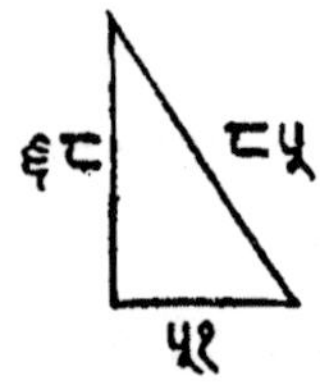

कर्णः। ८५। अयं द्विगुणः। १७०। द्विकेनेष्टेन हतः। ३४०। इष्ट। २। कृत्या। ४। सैकया। ५। भक्तो जाता कोटिः। ६८। इयमिष्टगुणा। १३६। कर्णो-। ६५। निता जातो भुजः। ५१।

चतुष्केणेष्टेन वा।

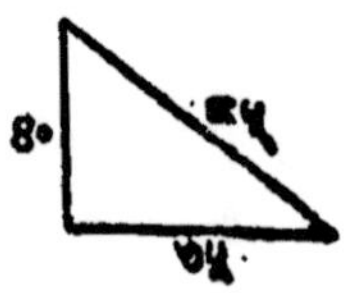

कोटिः। ४०।
भुजः। ७५।

प्रकारान्तरेण तत्करणसूत्रं वृत्तम्। इष्टवर्गेण सैकेन द्विघ्नः कर्णोऽथ वा हृतः। फलोन श्रवणः कोटिः फलमिष्टगुणं भुजः ॥ १४४ ॥

पूर्वोदाहरणे।

न्यासः

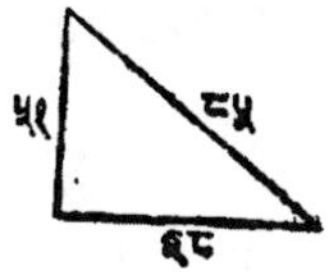

कर्णः। ८५। अत्रद्विकेनेष्टेन जातौ किल कोटिभुजौ। ५१। ६८।

चतुष्केण वा।

कोटिः। ७५। भुजः। ४०। अत्र दोःकोट्योर्नामभेद एव केवलं न स्वरूपभेदः।

अथेष्टाभ्यां भुजकोटिकर्णानयने करणसूत्रं वृत्तम्। इष्टयोराहतिर्द्विघ्नीकोटिर्वर्गान्तरं भुजः। कृतियोगस्तयोरेवं कर्णश्चाकरणीगतः ॥ १४५ ॥

उदाहरणम्। यैर्यैस्त्र्यस्रं भवेज्जात्यं कोटिदोःश्रवणैः सखे। त्रीनप्यविदितानेतान् क्षिप्रं ब्रूहि विचक्षण ॥ १४६ ॥

न्यासः

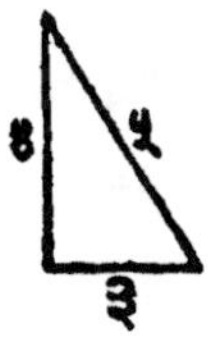

अत्रेष्टे। २। १। आभ्यां कोटिभुजकर्णाः। ४। ३। ५।

अथवेष्टे। २। ३। आभ्यां कोटिभुजकर्णाः। १२। ५। १३

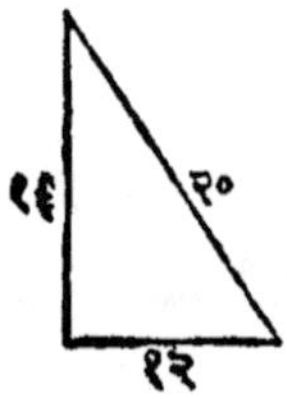

अथवेष्टे। २ । ४ । आभ्यां कोटिभुजकर्णाः । १६ । १२ । २० । एवमन्यत्रा- नेकधा।

कर्णकोटियुतौ भुजे च ज्ञाते पृथक्करणसूत्रं वृत्तम्। वंशाग्रमूला- न्तरभूमिवर्गो वंशोद्धृतस्तेन पृथग्युतोनौ। वंशौ तदर्द्धे भवतः क्रमेण वंशस्य खण्डे श्रुतिकोटिरूपे॥ १४७॥

उदाहरणम्। यदि समभुवि वेणुर्द्वित्रिपाणिप्रमाणो गणक पवनवेगादेकदेशे स भग्नः। भुवि नृपमितहस्तेष्वङ्गलग्नं तदग्रं कथय कतिषु मूलादेषभग्नः करेषु॥ १४८॥

न्यासः

वंशाग्र - मूलान्तर - भूमिर्भुजः। १६। वंशः कोटिकर्णयुतिः। ३२। जाते ऊर्द्ध्वाधःखण्डे। २०। १२।

बाहुकर्णयोगे दृष्टे कोट्यां च ज्ञातायां पृथक्करणसूत्रं वृत्तम्। स्तम्भस्य वर्गोऽहिविलान्तरेण भक्तः फलं व्यालविलान्तरालात्। शोध्यं तदर्द्धप्रमितैः करैः स्याद्विलाग्रतो व्यालकलापियोगः॥ १४९॥

उदाहरणम्। अस्ति स्तम्भतले विलं तदुपरि क्रीडाशिखण्डी स्थितः स्तम्भे हस्तनवोच्छ्रिते त्रिगुणितस्तम्भप्रमाणान्तरे। दृष्ट्वाहिं विलमात्रजन्तमपतत्तिर्यक् स तस्योपरि क्षिप्रं ब्रूहि तयोर्विलात्कतिमितैः साम्येन गत्योर्युतिः ॥ १५० ॥

न्यासः 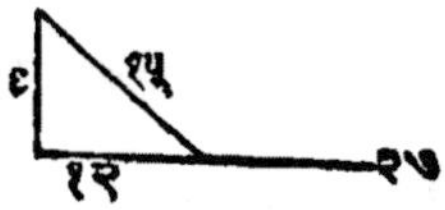

स्तम्भः। ९। अहिविलान्तरम्। २७। जाता विलयुत्योर्मध्यहस्ताः। १२।

कोटिकर्णान्तरे भुजे च दृष्टे पृथक्करणसूत्रं वृत्तम्। भुजाद्वर्गितात्कोटिकर्णान्तराप्तं द्विधा कोटिकर्णान्तरेणोनयुक्तम्। तदर्द्धे क्रमात् कोटिकर्णौ भवेतामिदं धीमताऽऽवेद्य सर्वत्र योज्यम् ॥ १५१ ॥

सखे पद्मतन्मज्जनस्थानमध्यं भुजः कोटिकर्णान्तरं पद्मदृश्यम्। नलः कोटिरेतन्मितं स्याद्यदम्भोवदैवं समानीय पानीयमानम् ॥ १५२ ॥

उदाहरणम्। चक्रक्रौञ्चाकुलितसलिले क्वापि दृष्टं तडागे तोयादूर्ध्वं कमलकलिकाग्रं वितस्तिप्रमाणम्। मन्दं मन्दं चलितमनिलेनाहतं हस्तयुग्मे तस्मिन्मग्नं गणक कथय क्षिप्रमम्भःप्रमाणम् ॥ १५३ ॥

न्यासः

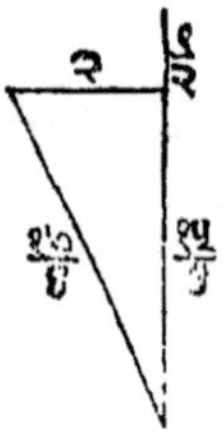

कोटिकर्णान्तरम् १/२ भुजः । २ । लब्धं जलगाम्भीर्यम् १५/४ इयं कोटिः इयमेव कलिकामानयुताजातः कर्णः १७/४

कोट्येकदेशेन युते कर्णे भुजे च दृष्टे कोटिकर्णज्ञानाय करणसूत्रं वृत्तम् । द्विनिघ्नतालोच्छ्रितिसंयुतं यत्सरोऽन्तरं तेन विभाजितायाः । तालोच्छ्रितेस्तालसरोऽन्तरघ्ना उड्डीयमानं खलु लभ्यते तत् ॥ १५४ ॥

उदाहरणम् । वृक्षाद्धस्तशतोच्छ्रयाच्छतयुगे वापीं कपिः । कोऽप्यगादुत्तीर्याथ परो द्रुतं श्रुतिपथेनोड्डीयकिञ्चिद् द्रुमात् । जातैवं समता तयोर्यदि गतावुड्डीयमानं कियद्विद्वन चेत्सुपरिश्रमोऽस्ति गणिते क्षिप्रं तदाऽऽचक्ष्व मे ॥ १५५ ॥

न्यासः

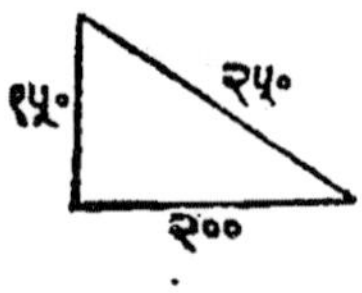

वृक्षवाप्यन्तरम् । २०० । वृक्षोच्छ्रायः । १०० । लब्धमुड्डीयमानम् । ५० । कोटिः ।१५०। कर्णः । २५० । भुजः । २०० ।

भुजकोट्योर्योगे कर्णे च ज्ञाते पृथक्करणसूत्रं वृत्तम् । कर्णस्य वर्गाद्द्विगुणाद्विशोधयेदोः कोटियोगः स्वगुणोऽस्य मूलम् ।

योगो द्विधा मूलविहीनयुक्तः स्यातां तदर्धे भुजकोटिमाने॥ १५६॥

उदाहरणम्। दशसप्ताधिकः कर्णस्त्र्यधिकाविंशतिः सखे। भुजकोटियुतिर्यत्र तत्र ते मे पृथग्वद॥ १५७॥

न्यासः

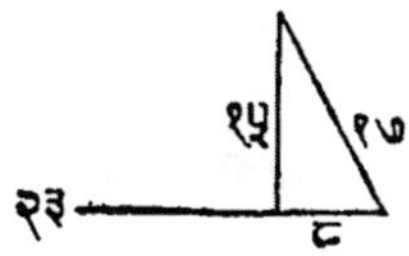

कर्णः। १७। दोः कोटियोगः। २३। जाते भुजकोटी। ८। १५।

उदाहरणम्। दोः कोट्योरन्तरं शैलाः कर्णो यत्र त्रयोदश। भुजकोटी पृथक्तत्र वदाशु गणकोत्तम॥ १५८॥

न्यासः

कर्णः। १३। भुजकोट्यन्तरम्। ७। लब्धे भुजकोटी। ५। १२।

लम्बावबाधाज्ञानाय करणसूत्रं वृत्तम्। अन्योऽन्यमूलाग्रगसूत्रयोगाद्वेण्वोर्वधे योगहृते च लम्बः। वंशौ स्वयोगेन हृतावभीष्टभुघ्नौ च लम्बोभयतः कुखण्डे॥ १५९॥

उदाहरणम्। पञ्चदशदशकरोच्छ्रायवेण्वोरज्ञात मध्यभूमिकयोः। इतरेतरमूलाग्रगसूत्रयुतेर्लम्बमानमाचक्ष्व॥ १६०॥

न्यासः

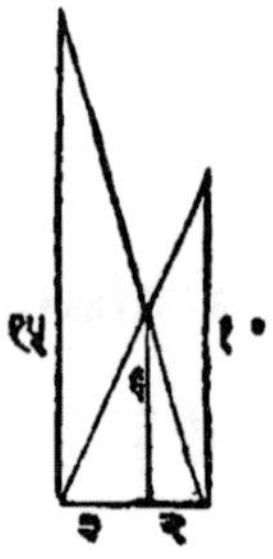

वंशौ। १५। १०। जातो लम्बः। ६। वंशान्तरभूः। ५। अत्र जाते भूखण्डे। ३। २। अथवा भूः। १०। खण्डे। ६। ४। वा भूः। १५। खण्डे। ९। ६। वा भूः। २०। खण्डे। १२। ८। एवं सर्वत्र लम्बः स एव।

यद्यत्र भूमितुल्ये भुजे वंशः कोटिस्तदा भूखण्डेन किमिति त्रैराशिकेन सर्वत्र प्रतीतिः।

अथ अक्षेत्रलक्षणे सूत्रम्। धृष्टोद्दिष्टमृजुभुजं क्षेत्रं यत्रैकबाहुतः स्वल्पा। तदितरभुजयुतिरथवा तुल्या ज्ञेयं तदक्षेत्रम्॥ १६१॥

उदाहरणम्। चतुरस्रे त्रिषड्द्वादर्कभुजास्त्र्यस्रे त्रिषट्नवाः। उद्दिष्टा यत्र धृष्टेन तदक्षेत्रं विनिर्दिशेत्॥ १६२॥

एते अनुपपन्ने क्षेत्रे।

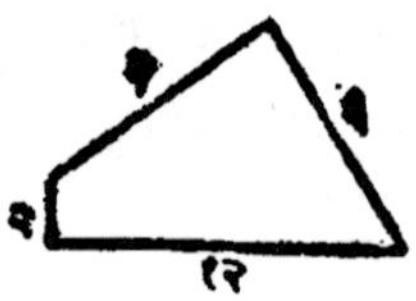

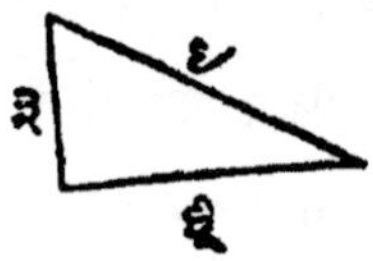

भुजप्रमाणा ऋजुशलाकाः भुजस्थानेषु विन्यस्यानुपपत्तिर्दर्शनीयेति।

अथबाधादिज्ञानाय करणसूत्रमार्य्याद्वयम्। त्रिभुजे भुजयोर्योगस्तदन्तरगुणो भुवा हृतो लब्ध्या। द्विःस्थाभूरूनयुता दलिताऽऽबाधे तयोः स्याताम्॥ स्वाऽऽबाधाभुजकृत्योरन्तरमूलं प्रजायते लम्बः। लम्बगुणं भूम्यर्द्धं स्पष्टं त्रिभुजे फलं भवति॥ १६३-१६४॥

उदाहरणम्। क्षेत्रे मही मनुमिता त्रिभुजे भुजौ तु यत्र त्रयोदशतिथिप्रमितौ च मित्र। तत्रावलम्बकमितिं कथयावबाधे क्षिप्रं तथा च समकोष्ठमितिं फलाख्याम्॥ १६५॥

न्यासः

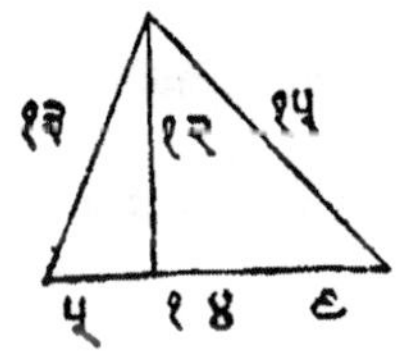

भूः। १४। भुजौ। १३। १५। लब्धे आबाधे। ५। ९। लम्बः। १२। क्षेत्रफलञ्च। ८४।

ऋणाऽऽबाधोदाहरणम्। दशसप्तदशप्रमौ भुजौ त्रिभुजे यत्र नवप्रमा मही। अबधे वद लम्बकं तथा गणितं गाणितिकाशु तत्र मे॥ १६६॥

न्यासः

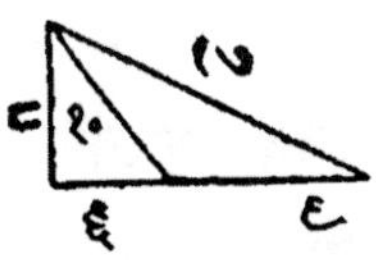

भुजौ। १०। १७। भूमिः। ९।

अत्र त्रिभुजे भुजयोर्योग इत्यादिना लब्धम्। २१। अनेन

भूरूना न स्यात् अस्मादेव भूरपनीता शेषार्द्धमृणगताऽऽबाधा दिग्वैपरीत्येनेत्यर्थः। तथा जाते आबाधे। ६। १५। अत उभयत्रापि जातो लम्बः। ८। फलम्। ३६।

चतुर्भुजत्रिभुजयोरस्पष्टफलानयने करणसूत्रं वृत्तम्। सर्वदोर्युतिदलं चतुःस्थितं बाहुभिर्विरहितञ्च तद्वधात्। मूलमस्फुटफलं चतुर्भुजे स्पष्टमेवमुदितं त्रिबाहुके ॥ १६७ ॥

उदाहरणम्। भूमिश्चतुर्दशमिता मुखमङ्कसङ्ख्यं बाहू त्रयोदशदिवाकरसम्मितौ च। लम्बोऽपि यत्र रविसंज्ञक एव तत्र क्षेत्रे फलं कथय तत्कथितं यदाद्यैः ॥ १६८ ॥

न्यासः

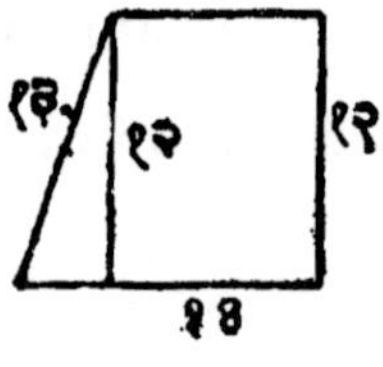

भूमिः। १४। मुखं। ९।
बाहू। १३। १२।
लम्बः। १२।

उक्तवत्करणेन जातं क्षेत्रफलं करणी। १९८००। अस्याः पदं किञ्चिन्न्यूनमेकचत्वारिंशच्छतम्। १४१। इदमत्र क्षेत्रे न वास्तवं फलं किन्तु लम्बेन निघ्नं कुमुखैक्यखण्डमित्यादिवक्ष्यमाणकरणेन वास्तवं फलम्। १३८।

अत्र त्रिभुजस्य पूर्वोदाहृतस्य।

न्यासः 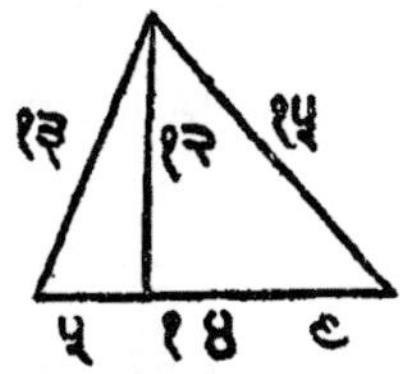

भूमिः। १४। भुजौ।१३।१५।

अनेनापि प्रकारेण त्रिबाहुके तदेव वास्तवं फलम्। ८४। अत्र चतुर्भुजस्यास्पष्टमुदितम्।

अथ स्थूलत्वनिरूपणार्थं सूत्रं सार्द्धवृत्तम्। चतुर्भुजस्यानियतौ हि कर्णौ कथं ततोऽस्मिन्नियतं फलं स्यात्। प्रसाधितौ तच्छ्रवणौ यदाद्यैः स्वकल्पितौ तावितरत्र न स्तः॥ तेष्वेव बाहुष्वपरौ च कर्णावनेकधा क्षेत्रफलं ततश्च॥ १६९-१७०॥

चतुर्भुजे हि एकान्तरकोणावाक्रम्याऽन्तः प्रवेश्यमानौ स्वसंसक्तं कर्णं सङ्कोचयतः इतरौ बहिरपसरन्तौ कोणौ स्वसंसक्तं कर्णं वर्द्धयतः अत उक्तान्तेष्वेव बाहुष्वपरौ च कर्णाविति।

लम्बयोः कर्णयोर्वैकमनिर्दिश्यापरान् कथं। पृच्छत्यनियतत्वेऽपि नियतश्चापि तत्फलम्॥ १७१॥

स पृच्छकः पिशाचो वा वक्ता वा नितरान्ततः। यो न वेत्ति चतुर्बाहुक्षेत्रस्यानियतां स्थितिम्॥ १७२॥

समचतुर्भुजाऽऽयतयोः फलानयने करणसूत्रं सार्द्धश्लोकद्वयम्।
इष्टा श्रुतिस्तुल्यचतुर्भुजस्य कल्प्याथ तद्वर्गविवर्जिता या। चतुर्गुणा बाहुकृतिस्तदीयं मूलं द्वितीयश्रवणप्रमाणम् ॥ १७३ ॥

अतुल्यकर्णाभिहतिर्द्विभक्ता फलं स्फुटं तुल्यचतुर्भुजे स्यात्। समश्रुतौ तुल्यचतुर्भुजे च तथाऽऽयते तद्भुजकोटिघातः ॥ १७४ ॥

चतुर्भुजेऽन्यत्र समानलम्बे लम्बेन निघ्नङ्कुमुखैक्यखण्डम् ॥ ॥ १७५ ॥

अत्रोद्देशकः। क्षेत्रस्य पञ्चकृतितुल्यचतुर्भुजस्य कर्णौ ततश्च गणितं गणक प्रचक्ष्व। तुल्यश्रुतेश्च खलु तस्य तथायतस्य यद्विस्तृतौ रसमिताष्टमितं च दैर्घ्यम् ॥ १७६ ॥

प्रथमोदाहरणे

न्यासः 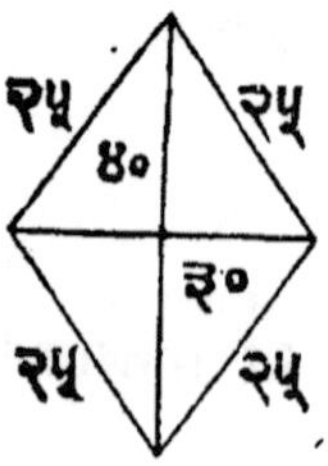

भुजाः। २५। २५। २५। २५।

अत्र त्रिंशन्मिता। ३०। मेकां श्रुतिं प्रकल्प्य जाताऽन्या। ४०। फलञ्च। ६००।

न्यासः 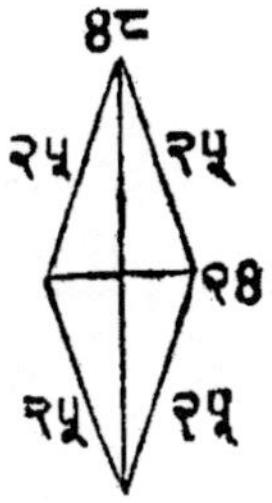

चतुर्दशमितामिकां। १४। श्रुतिं प्रकल्प्योक्तवत्करणेन जाताऽन्या श्रुतिः। ४८। फलञ्च। ३३६।

द्वितीयोदाहरणे

न्यासः 

तत्कृत्योर्योगपदं कर्णं इति जाता करणीगता श्रुतिरुभयत्र तुल्यैव। १२५०। गणितञ्च। ६२५।

अथायतस्य

न्यासः 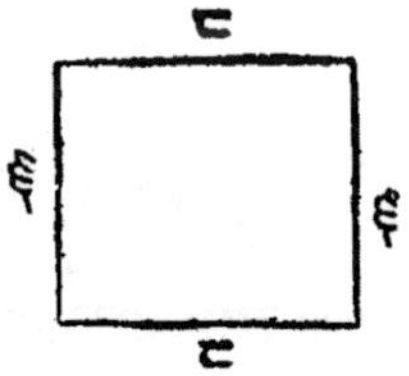

विस्तृतिः। ६। दैर्घ्यम्। ८। अस्य गणितम्। ४८।

उदाहरणम्। क्षेत्रस्य यस्य वदनं मदनारितुल्यं विश्वम्भरा द्विगुणितेन मुखेन तुल्या। बाहू त्रयोदशनखप्रमितौ च लम्बः सूर्योन्मितश्च गणितं वद तत्र किं स्यात्॥ १७७॥

न्यासः

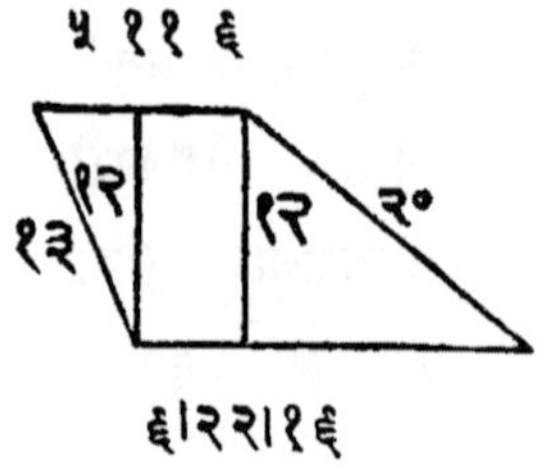

वदनम्। ११।
विश्वम्भरा। २२।
बाहू। १३। २०।
लम्बः। १२।

अत्र सर्वदोर्युतिदलमित्यादिना स्थूलं फलं। २५०। वास्तवन्तु लम्बेन निघ्नं कुमुखैक्यखण्डमिति जातं फलम्। १९८। क्षेत्रस्य खण्डत्रयं कृत्वा फलानि पृथगानीय ऐक्यं कृत्वास्य फलोपपत्तिर्दर्शनीया।

खण्डत्रयदर्शनम्।

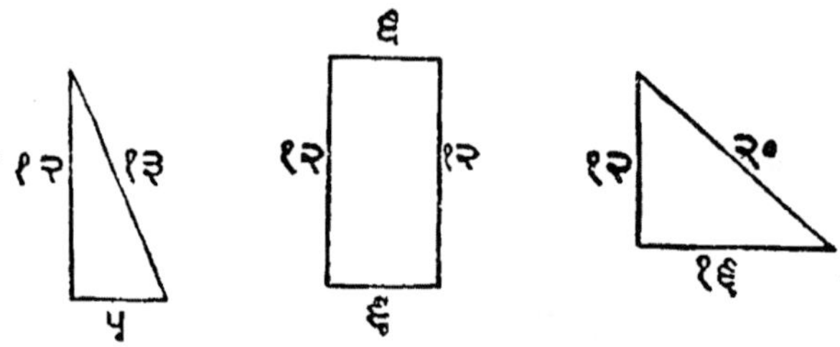

प्रथमस्य भुजकोटिकर्णाः। ५। १२। १३। द्वितीयस्यायतस्य विस्तृतिः। ६। दैर्घ्यम्। १२। तृतीयस्य भुजकोटिकर्णाः। १६। १२। २०।

अत्र त्रिभुजयोः क्षेत्रयोर्भुजकोटिघातार्द्धं फलं आयते चतुरस्रे क्षेत्रे तद्भुजकोटिघातः फलं यथा प्रथमक्षेत्रे फलम्। ३०। द्वितीये। ७२। तृतीये। ९६। एषामैक्यं सर्वक्षेत्रफलम्। १९८।

अथान्यदुदाहरणम्। पञ्चाशदेकसहिता वदनं यदीयं भूः पञ्चसप्ततिमिता प्रमितोऽष्टषष्ट्या। सव्यो भुजो द्विगुणविंशतिसम्मितोऽन्यस्तस्मिन् फलं श्रवणलम्बमिती प्रचक्ष्व॥ १७८॥

न्यासः 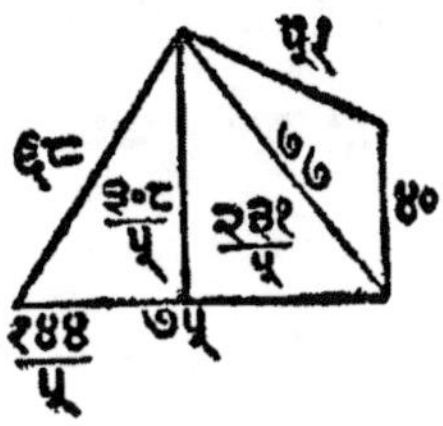

वदनम्। ५१।
भूमिः। ७५।
भुजौ। ६८। ४०।

अत्र फलावलम्बश्रुतीनां सूत्रं वृत्तार्द्धम्। ज्ञातेऽवलम्बे श्रवणः श्रुतौ तु लम्बः फलं स्यान्नियतन्तु तत्र॥ १७९॥

कर्णस्यानियतत्वाल्लम्बोऽप्यनियत इत्यर्थः। लम्बज्ञानाय करणसूत्रं वृत्तार्द्धम्। चतुर्भुजान्तस्त्रिभुजेऽवलम्बः प्राग्वद्भुजौ कर्णभुजौ मही भूः॥ १७९ऽऽ॥

अत्र लम्बज्ञानार्थं सव्यभुजाग्राद्दक्षिणभुजमूलगामी इष्टः कर्णः सप्तसप्ततिमितः कल्पितस्तेन चतुर्भुजान्तस्त्रिभुजं कल्पितं तत्राऽसौ कर्ण एको भुजः। ७७। द्वितीयस्तु सव्यभुजः। ६८। भूः सैव। ७५। अत्र प्राग्वल्लब्धो लम्बः ३०८/५

लम्बे ज्ञाते कर्णज्ञानार्थं सूत्रं वृत्तम्। यल्लम्बलम्बाश्रितबाहु-वर्गविश्लेषमूलं कथिताऽवधा सा। तदूनभूवर्गसमन्वितस्य यल्लम्ब-वर्गस्य पदं स कर्णः॥ १८० ॥

अत्र सव्यभुजाग्राल्लम्बः किल कल्पितः $\frac{३०८}{५}$ अतो जाताऽऽबाधा $\frac{१४४}{५}$ तदूनभूवर्गसमन्वितेत्यादिना जातः कर्णः। ७७।

द्वितीयज्ञानार्थं सूत्रं वृत्तद्वयम्। इष्टोऽत्र कर्णः प्रथमं प्रकल्प्य-स्त्र्यस्रे तु कर्णोभयतः स्थिते ये। कर्णं तयोः क्ष्मामितरौ च बाहू प्रकल्प्य लम्बाववधे च साध्ये॥ आबाधयोरेककुप्स्थयोर्यत् स्यादन्तरं तत्कृतिसंयुतस्य। लम्बैक्यवर्गस्य पदं द्वितीयः कर्णो भवेत् सर्वचतुर्भुजेषु॥ १८१-१८२॥

न्यासः

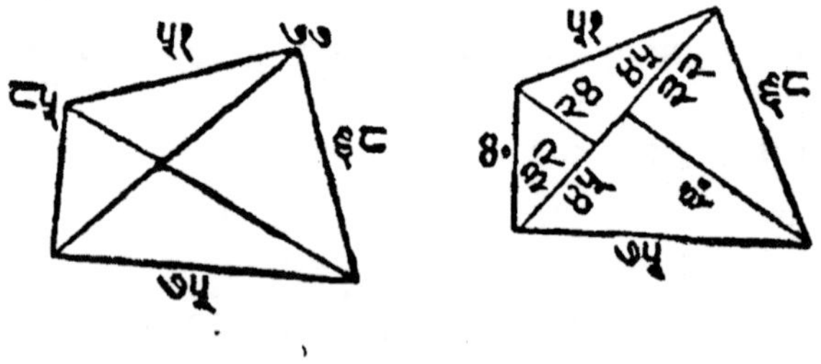

अत्र चतुर्भुजे सव्यभुजाग्राद्दक्षिणभुजमूलगामिनः कर्णस्य मानं कल्पितम्। ७७। तत्कर्णरेखावच्छिन्नस्य क्षेत्रस्य मध्ये कर्णरेखो-भयतो ये त्र्यस्रे उत्पन्ने तयोः कर्णं भूमिं तदितरौ च भुजौ प्रकल्प्य

प्राग्वल्लम्ब: आबाधा च साधिता। तद्दर्शनं लम्ब: । ६०। द्वितीयलम्ब: । २४। आबाधयो। ४५। २२। रेकककुप्स्थयोरन्तरस्य। १३। कृते। १६९। लम्बैक्य। ८४। कृतेश्च। ७०५६। योग: । ७२२५। तस्य पदं द्वितीयकर्णप्रमाणम्। ८५।

अत्रेष्टकर्णकल्पने विशेषोक्तिसूत्रं सार्द्धवृत्तम्।

कर्णाश्रितस्वल्पभुजैक्यमुर्वीं प्रकल्प्य तच्छेषभुजौ च बाहू।
साध्योऽवलम्बोऽथ तथान्यकर्ण: स्वोर्व्या: कथं चित्क्रियतोन दीर्घ: ॥
तदन्यलम्बान्न लघुस्तथेदं ज्ञात्वेष्टकर्ण: सुधिया प्रकल्प्य: ॥ १८३ ॥

चतुर्भुजो हि एकान्तरकोणयोराक्रम्य सङ्कोच्यमान स्त्रिभुजत्वं याति तत्रैककोणलग्नलघुभुजयोरैक्यं भमिरितरौ भुजौ च। तल्लम्बादून: सङ्कोच्यमान: कर्ण: कथञ्चिदपि न स्यात् तदितरो भूमेरधिको न स्यादेवमुभयथापि। एतदनुक्तमपि बुद्धिमता ज्ञायते।

विषमचतुर्भुजफलानयनाय करणसूत्रं वृत्तार्द्धम्। त्र्यस्रे तु कर्णोभयत: स्थिते ये तयो: फलैक्यं फलमत्र नूनम् ॥ २८४ ॥

अनन्तरोक्तक्षेत्रान्तस्त्र्यस्रयो: फले। ८२४। १३१०। अनयोरैक्यं। ३२३४। तस्य फलम्।

समानलम्बस्याबधादिज्ञानाय करणसूत्रं वृत्तद्वयम्। समानलम्बस्य चतुर्भुजस्य मुखोनभूमिं परिकल्प्य भूमिम्। भुजौ भुजौ त्र्यस्रवदेव साध्ये तस्यावधे लम्बमितिस्ततश्च॥ आबाधयोना चतुरस्रभूमिस्तल्लम्बवर्गैक्यपदं श्रुतिः स्यात्। असमानलम्बे लघुदोःकुयोगान्मुखान्यदोःसंयुतिरल्पिका स्यात्॥ १८५—१८६॥

उदाहरणम्। द्विपञ्चाशन्मितत्र्येकचत्वारिंशन्मितौ भुजौ। मुखन्तु पञ्चविंशत्या तुल्यं षष्ट्या मही किल॥ अतुल्यलम्बकं क्षेत्रमिदं पूर्वैरुदाहृतम्। षट्पञ्चाशत्त्रिषष्टिश्च नियते कर्णयोर्मिती॥ कर्णौ तत्राऽपरौ ब्रूहि समलम्बञ्च तच्छ्रुती॥ १८७—१८८॥

न्यासः

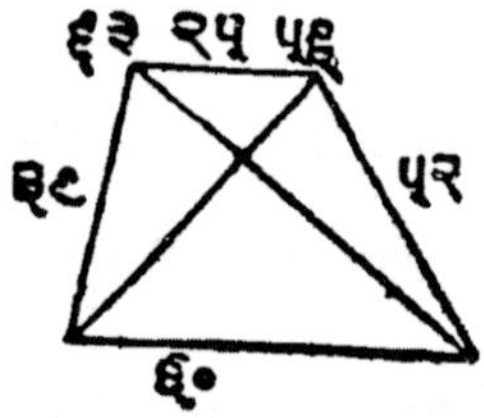

अत्र बृहत्कर्णं त्रिषष्टिमिति प्रकल्प्यात् जातः प्राग्वदन्यः कर्णः। ५६।

अथ षट्पञ्चाशत्स्थाने द्वात्रिंशन्मितं कर्णं। २२। प्रकल्प्य

न्यासः

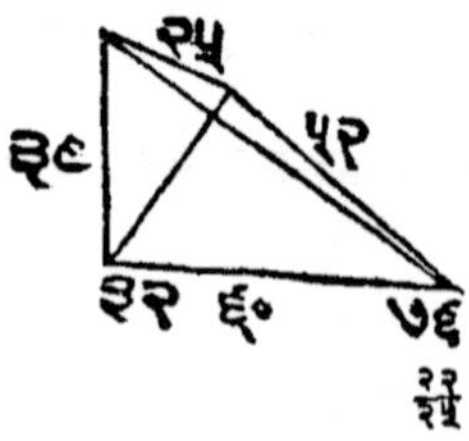

प्राग्वत्साध्यमाने कर्णे जातं करणीखण्डद्वयम् । ६२१ । । २७०० । अनयोर्मूलयो- २४ $\frac{23}{25}$ ५१ $\frac{24}{25}$ रैक्यं द्वितीयः कर्णः ७६ $\frac{22}{25}$

अथ तदेव क्षेत्रं क्षेत् समलम्बम् ।

न्यासः

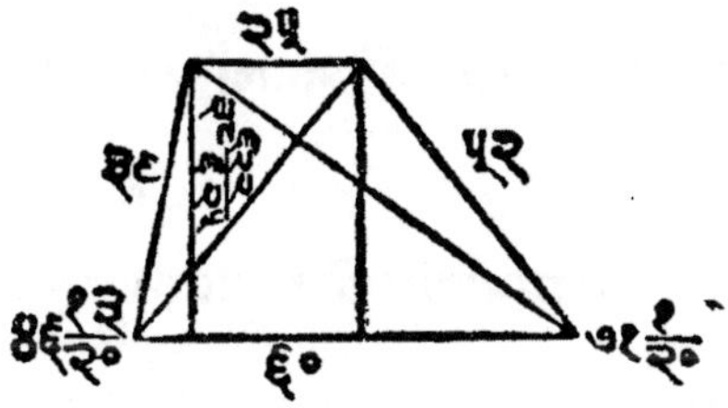

तदा मुखोन भूमिं परिकल्प्य भूमिमिति ज्ञानार्थं त्र्यस्रं कल्पितम् ।

न्यासः

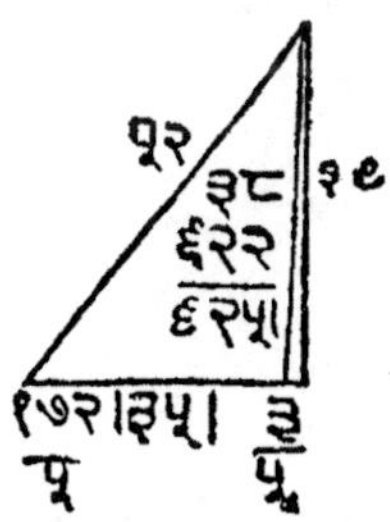

अत आबाधे जाते $\frac{3}{5}$ $\frac{172}{5}$

लम्बश्च करणीगतो जातः $\frac{38016}{25}$

आसन्नमूलकरणेन जातः ३८ $\frac{622}{625}$

अयं तत्र चतुर्भुजे समलम्बः। लघ्वाऽऽबाधोनितसमभूमेर्लम्बस्य च वर्गयोगः। ५०४९। अयं कर्णवर्गः। एवं बृहदाबाधातो द्वितीयकर्णवर्गः। २१७६। अनयोरासन्नमूलकरणेन जातौ कर्णौ

७१ ४६
१/२० १३/२०

एवं चतुरस्रे तेष्वेव बाहुष्वन्यौ कर्णौ बहुधा भवतः। एवमनियतत्वेऽपि नियतावेव कर्णावानीतौ ब्रह्मगुप्ताद्यैस्तादानयनं यथा।

कर्णाश्रितभुजघातैक्यमुभयथाऽन्योन्यभाजितं गुणयेत्। योगेन भुजप्रतिभुजहत्योः कर्णौ पदे विषमे॥ १९०॥

न्यासः

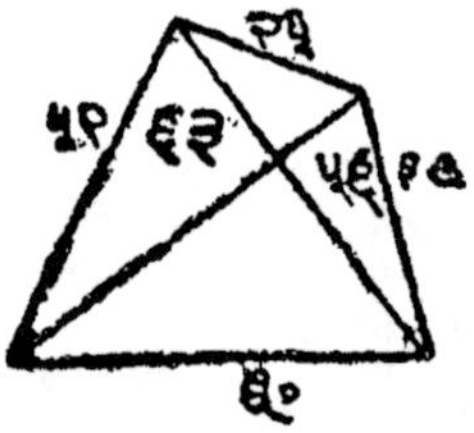

कर्णाश्रितभुजघातेति एकवारमनयोः। २५। ३९। घातः। ९७५। तथा। ५२। ६०। अनयोर्घातः। ३१२०। घातयोर्द्वयोरैक्यम्। ४०९५। तथा द्वितीयवारम्। २५। ५२। अनयोर्घातो जातम्। १३००। तथा द्वितीयवारम्। ३९। ६०।

अनयोर्घाते । २२४० । घातयोर्द्वयोरैक्यम् । ३६४० । भुज-प्रतिभुज-। ५२ । ३९ । घातः । २०२८ । पश्चात् । २५ । ६० । अनयोर्बधः । १५०० । तयोरैक्यम् । ३५२८ । अनेनैक्येन । । ३६४० । गुणितं जातं पूर्वैक्यम् । १२८४१९२० । प्रथम-कर्णाश्रितभुजघातैक्येन । ४०९५ । भक्तं लब्धम् । ३१३६ । अस्य मूलम् । ५६ । एकः कर्णः । तथा द्वितीयकर्णार्थ-प्रथमकर्णाश्रितभुजघातैक्यम् । ४०९५ । भुजप्रतिभुजबधयोग- । । ३५२८ । गुणितं जातम् । १४४४७१६० । अन्यकर्णा-श्रितघातैखेन :। ३६४० । भक्तं लब्धम् । ३९६९ । अस्य मूलम् । ६३ । द्वितीयः कर्णः । अस्य कर्णानयनस्य प्रक्रिया-गौरवं लघुप्रक्रियादर्शनद्वारेणाह । अभीष्टजात्यद्वयबाहुकोटयः परस्परं कर्णहताभुजा इति । चतुर्भुजं यद्विषमं प्रकल्पितं श्रुती तु तत्र त्रिभुजद्वयात्ततः ॥ बाह्वोर्बधः कोटिबधेन युक् स्यादेका श्रुतिः कोटिभुजाबधैक्यम् । अन्या लघौ सत्यपि साधनेऽस्मिन् पूर्वैः कृतं यद् बहु तन्न विद्मः ॥ १९१—१९२ ॥

जात्यक्षेत्रद्वयस्य

न्यासः

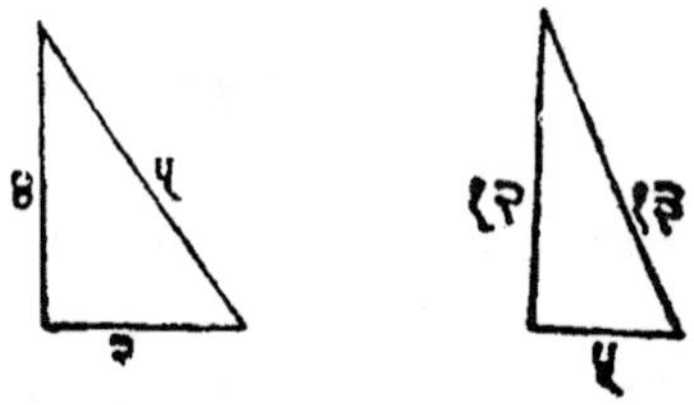

एतयोरितरेतरकर्णहताभुजाः कोटयः इतरेतरकर्णहताः कोटयो भुजा इति कृते जातम् । २५ । ६० । ५२ । ३९। तेषां महतो भूर्लघुमुखमितरौ बाहू इति प्रकल्प्य क्षेत्र-दर्शनं इमौ कर्णौ महतायासेनानोतौ। ६३। ५६। अस्यैव जात्यद्वयस्योत्तरोत्तरभुजकोव्योर्घातौ जातौ। ३६। २०। अनयो-रैक्यमेकः कर्णः। ५६। बाह्वोः। ३ । ५ । कोव्यो- । ४ । । १२ । ख घातौ । १५ । ४८ । अनयोरैक्यमन्यः कर्णः। ६३। एवं सुखेन श्रुती स्याताम्। अथ यदि पार्श्वभुजमुखयो-र्व्यत्ययं कृत्वा न्यस्तं क्षेत्रम्

न्यासः

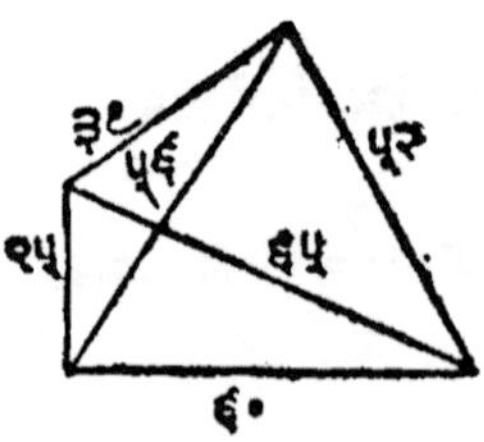

तदा जात्यद्वयकर्णयोर्वधः। ६५। द्वितीयकर्णः।

अथ सूचीक्षेत्रोदाहरणम्। क्षेत्रे यत्र शतत्रयं क्षितिमिति-स्तत्त्वेन्दु- (१२५) तुल्यं मुखं बाहू खोत्कृतिभिः (२६०) शरातिधृतिभि- (१९५) स्तुल्यौ च तत्र श्रुतिः। एका खाष्टयमैः (२८०) समा तिथिगुणै- (३१५) रन्याथ तल्लम्बकौ तुल्यौ गोधृतिभि- (१८९) स्तथाजिनयमै (२२४) र्योगाच्छ्रवोलम्बयोः॥ तत्खण्डे कथयाधरे श्रवणयोर्योगाच्च लम्बावधे तत् सूची निज-मार्गवृद्धभुजयोर्योगाद्यथा स्यात्ततः। साबाधं वत लम्बकद्व भुजयोः सूच्याः प्रमाणे च के। सर्वं गाणितिक प्रचक्ष्व नितरां क्षेत्रेऽत्र दक्षोऽसि चेत्॥ १८३—१८४॥

न्यासः

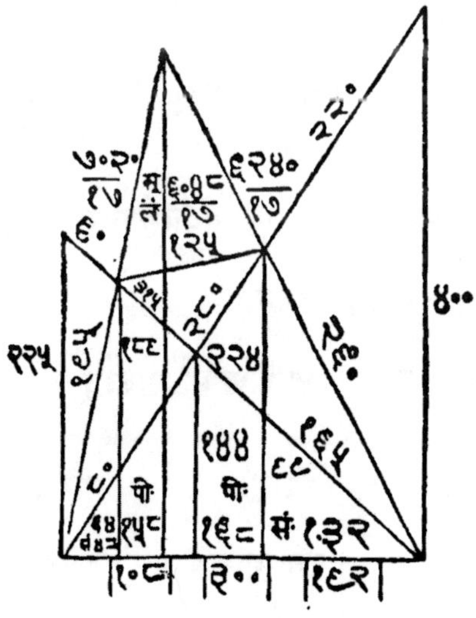

भूमानम्। ३०० । मुखम् । १२५ । बाहू । २६० । । १९५ । कर्णौ । २८० । ३१५ । लम्बौ । १८९ । । २२४ ।

अथ सन्ध्याद्यानयनाय करणसूत्रं आर्याद्वयम्। लम्बतदाश्रितबाह्वोर्मध्यं सन्धिरपरमस्य लम्बस्य। सन्ध्योना भूः पीठं साध्यं यस्याधरं खण्डम्॥ सन्धिर्द्विस्थः परलम्बश्रवणहतः परस्य पीठेन। भक्तो लम्बश्रुत्योर्योगात् स्यातामधःखण्डे॥ २९५—१९६॥

लम्बः। १८९ । तदाश्रितभुजः । १९५ । अनयोर्मध्ये यल्लम्बलम्बाश्रितबाहुवर्गेत्यादिनाऽऽगताऽऽबाधा सन्धिसंज्ञा। ४८ । तदूनितभूरिति द्वितीयाऽऽबाधा सा पीठसंज्ञा। २५२।

एवं द्वितीयलम्बः। २२४ । तदाश्रितभुजः। २६०। पूर्ववत् सन्धिः। १३२। पीठम्। १६८।

अथाद्यलम्बन १८९। स्याधःखण्डं साध्यम्।

अस्य सन्धिः। ४८। द्विस्थः। ४८ । परलम्बेन। २२४ श्रवणेन च । २८० । पृथग्गुणितः। १०७५२। १३४४०। परस्य पीठेन । १६८ । भक्तो लब्धं लम्बाधःखण्डम्। ६४। श्रवणाधःखण्डम् । ८०।

एवं द्वितीयलम्बस्य। २२४। सन्धिः। १३२। परलम्बेन। १८९। कर्णेन च। ३१५। पृथग्गुणितः परस्य पीठेन। २५२। भक्तो लब्धं लम्बाधःखण्डम्। ९९। भूवणाधःखण्डञ्च। १६५।

अथ कर्णयोर्योगादधोलम्बज्ञानार्थं सूत्रं वृत्तम्। लम्बौ भूघ्नौ निजनिजपीठविभक्तौ च वंशौ स्तः। ताभ्यां प्राग्वच्छ्रुत्योर्योगाल्लम्बः कुखण्डे च॥ १९७॥

लम्बौ। १८९। २२४। भू-। ३००। घ्नौ जातौ। ५६७००। ६७२००। स्वस्वपीठाभ्यां। २५२। १६८। भक्तौ एवमत्र लब्धौ वंशौ। २२५। ४००। आभ्याम् अन्योऽन्यमूलाग्रगसूत्रयोगादित्यादिकरणेन लब्धः कर्णयोगादधोलम्बः। १४४। भूखण्डे च। १०८। १९२।

अथ सूच्याऽऽबाधालम्बभुजज्ञानार्थं सूत्रं वृत्तत्रयम। लम्बहतो निजसन्धिः परलम्बगुणः समाह्वयो ज्ञेयः। समपरसन्ध्योरैक्यं हारस्तेनोद्धृतौ तौ च॥ समपरसन्धी भूघ्नौ सूच्याऽऽबाधे पृथक् स्याताम्। हारहृतः परलम्बः सूचीलम्बो भवेद्भूघ्नः॥ सूचीलम्बघ्नभुजौ निजनिजलम्बोद्धृतौ सूच्याः। एवं चेत्क्षोदः प्राज्ञैस्त्रैराशिकाज्ज्ञेयः॥ १९८—२००॥

अत्र किलायं लम्बः । २२४ । अस्य सन्धिः । १३२ । अयं परलम्बेन। १८८। गुणितो। २२४। ऽनेन भक्तो जातः समाह्वयः $\frac{८८२}{९}$

अस्य परसन्धेश्च । ४८ । योगो हारः $\frac{१२०५}{९}$

अनेन भू- । ३०० । घ्न समः $\frac{२६०३००}{९}$

परसन्धि $\frac{१४४००}{१}$ श्च भक्तो जाते सूच्याऽऽबाधे $\frac{३५६४}{१७}$ $\frac{१५३६}{१७}$

एवं द्वितीयसमाह्वयः $\frac{५१२}{९}$

द्वितीयो हारः $\frac{१५००}{९}$

अनेन भूघ्न स्त्रीयः समः $\frac{१५३००}{९}$ परसन्धिश्च $\frac{३१६००}{१}$ भक्तो जाते सूच्याऽऽबाधे $\frac{१५३६}{१७}$ $\frac{३५६४}{१७}$ परलम्बः । २२४ । भूमि । ३०० । गुणो हारेण $\frac{१७००}{९}$ भक्तो जातः सूचीलम्बः $\frac{६०४८}{१७}$

सूचीलम्बेन भुजौ । १८५ । २६० । गुणितौ स्वस्वलम्बाभ्याम्। १८८। २२४ । यथाक्रमं भक्तौ जातौ स्वमार्गवृद्धौ सूचीभुजौ $\frac{६२४०}{१७}$ $\frac{७०२०}{१७}$

एवमत्र सर्वत्र भागहारराशिं प्रमाणं गुण्यगुणकौ तु यथायोग्यं फलेच्छे प्रकल्प्य सधिया त्रैराशिकमूह्यम ।

अथ वृत्तक्षेत्रे करणसूत्रं वृत्तम्। व्यासे भनन्दाग्नि (३९२७) हते विभक्ते खबाणसूर्यैः (१२५०) परिधिः स सूक्ष्मः। द्वाविंशतिघ्ने विहृतेऽथ शैलैः (७) स्थूलोऽथवा स्याद्व्यवहारयोग्यः॥ २०१॥

उदाहरणम्। विष्कम्भमानं किल सप्त यत्र तत्र प्रमाणं परिधेः प्रचक्ष्व। द्वाविंशतिर्यत्परिधिप्रमाणं तद्व्याससङ्ख्यां च सखे विचिन्त्य॥ २०२॥

न्यासः

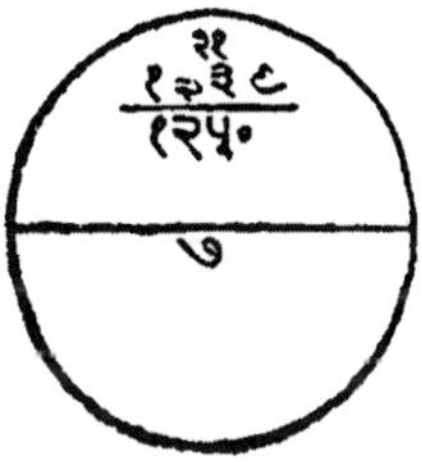

व्यासमानं। ७।

लब्धं परिधिमानं

२१ $\frac{१२३८}{१२५०}$

स्थूलो वा परिधिर्लब्धा। २२।

अथवा परिधियो व्यासानयनाय

न्यासः

गुणहारविपर्ययेण

व्यासमानं सूक्ष्मं ७ $\frac{११}{३९२७}$

स्थूलं वा। ७।

वृत्तगोलयोः फलानयने करणसूत्रं वृत्तम्। वृत्तक्षेत्रे परिधिगुणितव्यासपादः फलं यत् क्षुण्णं वेदैरुपरि परितः कन्दुकस्येव जालम्।

गोलस्यैवं तदपि च फलं पृष्ठजं व्यासनिघ्नं षड्भिर्भक्तं भवति नियतं गोलगर्भे घनाख्यम् ॥ २०३ ॥

उदाहरणम्। यद्व्यासस्तुरगैर्मितः किल फलं क्षेत्रे समे तत्र किं व्यासः सप्तमितश्च यस्य सुमते गोलस्य तस्यापि किम्। पृष्ठे कन्दुकजालसन्निभफलं गोलस्य तस्यापि किं मध्ये ब्रूहि घनं फलञ्च विमलाञ्चेद्वेत्सि लीलावतीम् ॥ २०४ ॥

वृत्तक्षेत्रफलदर्शनाय

व्यासः

व्यासः। ७।

परिधिः २१ $\frac{१२३९}{१२५०}$

क्षेत्रफलम् ३८ $\frac{२४२३}{५०००}$

गोलपृष्ठफलदर्शनाय

व्यासः

व्यासः। ७।

गोलपृष्ठफलम्। १५३ $\frac{११७३}{१२५०}$

गोलान्तर्गतघनफलदर्शनाय

न्यासः

न्यासः। ७। गोलस्यान्तर्गत घनफलम् १७९ १४८७/२५००

अथ प्रकारान्तरेण तत्फलानयने करणसूत्रं सार्द्धवृत्तम्। व्यासस्य वर्गे भनवाग्निनिघ्ने सूक्ष्मं फलं पञ्चसहस्रभक्ते। रुद्राहते शक्रहृतेऽथवा स्यात् स्थूलं फलं तद्व्यवहारयोग्यम्। घनीकृतव्यासदलं निजैकविंशांशयुग्गोलघनं फलं स्यात्॥ २०५-२०६॥

न्यासः। ७। अस्य वर्गे। ४९। भनवाग्निनिघ्ने पञ्चसहस्रभक्ते तदेव सूक्ष्मं फलम् ३८ ७४२३/५०००

अथवा व्यासस्य वर्गे। ४९। रुद्राहते। ५३९। शक्रहृते लब्धं स्थूलं फलम् ३८ १/२

घनीकृतव्यासदलं ३४३/२

निजैकविंशांशयुक् गोलस्य घनफलं स्थूलम् १७९ २/३

शरजीवानयनाय करणसूत्रं सार्धवृत्तम्। ज्याव्यासयोगान्तरघातमूलं व्यासस्तदूनो दलितः शरः स्यात्। व्यासाच्छरोनाच्छरसङ्गुणाच्च मूलं द्विनिघ्नं भवतीह जीवा। जीवार्धवर्गे शरभक्तयुक्ते व्यासप्रमाणं प्रवदन्ति वृत्ते ॥ २०६-१०७ ॥

उदाहरणम्। दशविस्तृतिवृत्तान्तर्यत्र ज्या षण्मिता सखे। तत्रेषुं वद बाणाज्ज्यां ज्याबाणाभ्यां च विस्तृतिम् ॥ २०८ ॥

व्यासः

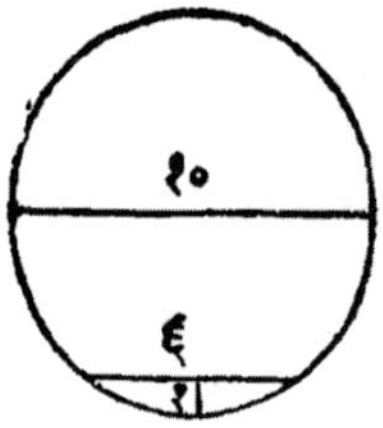

व्यासः। १०। ज्या। ६। योगः। १६। अन्तरं। ४। घातः। ६४।

मूलम्। ८। एतदूनो व्यासः २। दलितः। १। जातः शरः। १।

व्यासात्। १०। शरोनात्। ९। शर। १। सङ्गुणात्। ९। मूलम्। ३। द्विनिघ्नं जाता जीवा। ६।

एवं ज्ञाताभ्यां ज्याबाणाभ्यां व्यासानयनं यथा। जीवा-। ६। र्धं। ३। वर्गे। ९। शर। १। भक्ते। ९। शर। १। युक्ते जातो व्यासः। १०।

अथ वृत्तान्तस्त्र्यस्रादिनवास्रान्तक्षेत्राणां भुजमानानयनाय

करणसूत्रं वृत्तत्रयम् । त्रिद्व्यङ्काग्निनभश्चन्द्रे- (१०३९२३) स्त्रिबाणाष्टयुगाष्टभिः (८४८४३)। वेदाग्निबाणखाश्वैश्च (७०५३४) खखाभ्राभ्ररसैः (६००००) क्रमात् ॥ बाणेषुनखबाणैश्च (५२०५५) द्विद्विनन्देषुसागरैः (४५९२२)। कुरामदशवेदैश्च (४१०३१) वृत्तव्यासे समाहते ॥ खखखाभ्रार्क (१२००००) सम्भक्ते लभ्यन्ते क्रमशो भुजाः। वृत्तान्तस्त्र्यस्रपूर्वाणां नवास्रान्तं पृथक् पृथक् ॥ २०९-२११ ॥

उदाहरणम्। सहस्रद्वितयव्यासं यद्वृत्तं तस्य मध्यतः। समत्र्यस्रादिकानां मे भुजान वद पृथक् पृथक् ॥ २१२ ॥

अथ वृत्तान्तस्त्रिभुजे भुजमानानयनाय

न्यासः

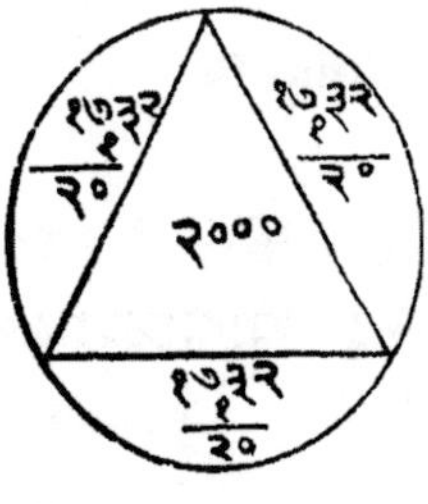

व्यासः । २००० । त्रिद्व्यङ्काग्निनभश्चन्द्रैः १०३९२३ । गुणितः । २०७८४६००० ।

खखखाभ्रार्कैः । १२०००० । भक्तो लब्धं त्र्यस्रे भुजमानम् १७३२ $\frac{१}{२०}$

वृत्तान्तश्चतुर्भुजे भुजमानानयनाय

व्यासः

व्यासः। २०००। त्रिबाणाष्टयुगाष्टभिः । ८४८५३ । गुणितः। १६९७०६००० । खखखाभ्रार्कैः। १२००० ।

भक्ते लब्धं चतुरस्रे भुजमानम् १४१४ १३/६०

वृत्तान्तःपञ्चभुजे भुजमानानयनाय

व्यासः

व्यासः। १०००। वेदाग्निबाणखाश्वैः। ७०५३४ । गुणितः। १४१०६८००० । खखखाभ्रार्कैः।१२००००।

भक्ते लब्धं पञ्चास्रे भुजमानम् ११७५ १७/३०

वृत्तान्तःषड्भुजे भुजमानानयनाय

व्यासः

व्यासः। २०००। खखाभ्ररसैः। ६००००। गुणितः। १२००००००० । खखखाभ्रार्कैः। १२००० । भक्ते लब्धं षड्स्रे भुजमानम्। १००००।

वृत्तान्तःसप्तभुजे भुजमानानयनाय

न्यासः

व्यासः । २००० ।
बाणेषुनखबाणैः ।
५२०५५ । गुणितः ।
। १०४११०००० ।
खखखाभ्राकैः ।
१२०००० । भक्ते
लब्धं सप्तास्रे

भुजमानम् ८६७ $\frac{७}{१२}$

वृत्तान्तरष्टभुजे भुजमानानयनाय

न्यासः

व्यासः । २००० ।
द्विद्विनन्देषुसागरैः ।
४५९२२ । गुणितः ।
९१८४४००० ।: खख-
खाभ्राकैः ।:१२०००० ।
भक्ते लब्धमष्टास्रे

भुजमानम् ७६५ $\frac{११}{३०}$

वृत्तान्तर्नवभुजे भुजमानानयनाय

न्यासः

व्यासः। २००० । कुरामदग्निवेदैः। ४१०३१ । गुणितः। ८२०६२००० । खखखाभ्रार्कैः। १२०००० । भक्ते लब्धं नवास्रे

भुजमानम् ६८३ $\frac{१७}{२०}$

एवमिष्टव्यासादेभ्योऽन्या अपि जीवाः सिद्ध्यन्तीति तास्तु गोले ज्योत्पत्तौ वक्ष्ये।

अथ स्थूलजीवाज्ञानार्थं लघुक्रियया करणसूत्रं वृत्तम्। चापोननिघ्नपरिधिः प्रथमाह्वयः स्यात् पञ्चाहतः परिधिवर्गचतुर्थभागः। आद्योनितेन खलु तेन भजेच्चतुर्घ्नव्यासाहतं प्रथममाप्तमिह ज्यका स्यात्॥ २१३॥

उदाहरणम्। अष्टादशांशेन वृतेः समानमेकादिनिघ्नेन च

यत्र चापम्। पृथक् पृथक् तत्र वदाशु जीवां खार्कैर्मितं व्यासदलञ्च यत्र ॥ २१४ ॥

न्यासः

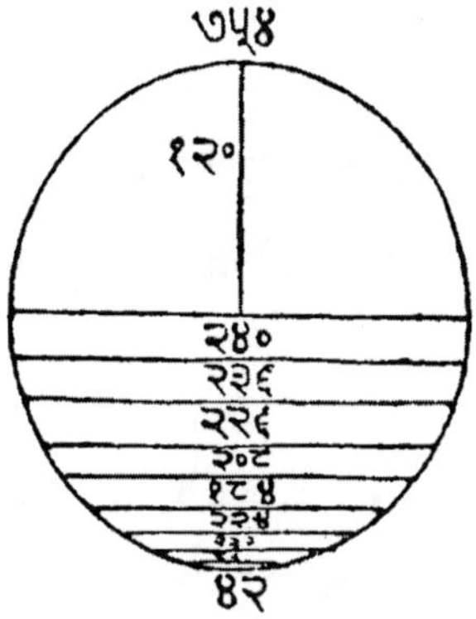

व्यासः। २४०। अत्र किलाङ्कलाघवाय विंशतेः सार्द्धार्कशतांशमिलितः सूक्ष्मपरिधिः। ७५४। अस्याष्टादशांशः। ४२। अत्राप्यङ्कलाघवाय द्वयोरष्टादशांशयुतो गृहीतः। अनेन पृथक् पृथगेकादिगुणितेन तुल्ये धनुषि कल्पिते ज्याः साध्याः।

अथवाऽत्र सुखार्थं परिधिरष्टादशांशेन परिधिं धनूंषि चापवच्च ज्याः साध्यास्तथाऽपि ता एव भवन्ति।

अपवर्त्तिते न्यासः। परिधिः। १८। चापानि च। १। । २। ३। ४। ५। ६। ७। ८। ९। यथोक्तकरणेन लब्धा जीवाः। ४२। ८२। १२०। १५४। १८४। २०८। २२६। २३६। २४०।

अथ चापानयनाय करणसूत्रं वृत्तम्। व्यासाब्धिघातयुतमौर्विकया विभक्तो जीवाङ्घ्रिपञ्च गुणितः परिधेस्तु वर्गः। लब्धोनितात्परिधिवर्गचतुर्थभागादाप्ते पदे वृतिदलात् पतिते धनुः स्यात् ॥ २१५ ॥

उदाहरणम्। विहिता इह ये गुणास्ततो वद तेषा मधुना धनुर्मितिम्। यदि तेऽस्ति धनुर्गुणक्रियागणिते गणितिकातिनैपुणम् ॥ २१६ ॥

न्यासः। ४२। ८२। १२०। १५४। १८४। २०८। २२६। २३६। २४०। स एवापवर्त्तितपरिधिः। १८।

जीवाङ्घ्रिणा ²¹/₄ पञ्चभि-। ५। श्च परिधे-। १८। र्वर्गो। ३२४। गुणितः। १७०१०। व्यासा-। २४०। ब्धि। ४। घात। ९६०। युतमौर्विकयानया। १००२। विभक्तो लब्धः। १७। अत्राङ्कलाघवाय चतुर्विंशतेर्द्व्यधिकसहस्रांशयुतो गृहीतोऽनेनोनितात्परिधि। १८। वर्ग। ३२४। चतुर्थभागात्। ६४। पदे प्राप्ते। ८। वृति। १८। दलात्। ९। पतिते। १। जातं धनुः। एवं जातानि धनूंषि। १। २। ३। ४। ५। ६। ७। ८। ९। एतानि परिधेरष्टादशांशेन गुणितानि स्युः।

इति चैत्रव्यवहारः।

[सप्तमोऽध्यायः।]

अथ खातव्यवहारे करणसूत्रं सार्द्धार्य्या। गणयित्वा विस्तारं बहुषु स्थानेषु तद्युतिर्भाज्या। स्थानकमित्या सममितिरेवं दैर्घ्ये च वेधे च॥ क्षेत्रफलं वेधगुणं खाते घनहस्तसङ्ख्या स्यात्॥ २१७-२१८॥

उदाहरणम्। भुजवक्रतया दैर्घ्यं दशेशार्ककरैर्मितम्। त्रिषु स्थानेषु षट्पञ्चसप्तहस्ता च विस्तृतिः। यस्य खातस्य वेधोऽपि द्विचतुस्त्रिकरः सखे। तत्र खाते कियन्तः स्युर्घनहस्ता प्रचक्ष्व मे॥ २१९-२२०॥

न्यासः

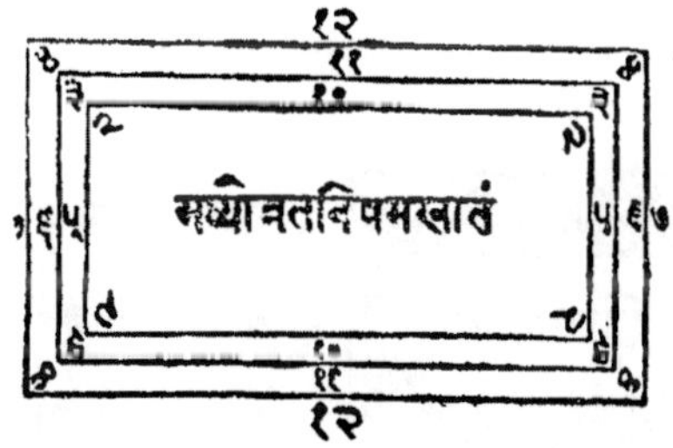

अत्र सममितिकरणेन विस्तारे हस्ताः। ६। दैर्घ्ये। ११। वेधे। ३।

तत् क्षेत्रदर्शनं यथा

न्यासः

यथोक्तकरणेन लब्धा घनहस्तसङ्ख्या। १९८।

खातान्तरे करणसूत्रं सार्द्धवृत्तम्। मुखजतलजतद्युतिजक्षेत्रफलैक्यं हृतं षड्भिः। क्षेत्रफलं सममेतद्वेधगुणं घनफलं स्पष्टम्। समखातफलत्र्यंशः सूचीखाते फलं भवति॥ २२१॥

उदाहरणम्। मुखे दशद्वादशहस्ततुल्यं विस्तारदैर्घ्यन्तु तले तदर्द्धम्। यस्याः सखे सप्तकरश्च वेधः का खातसङ्ख्या वद तत्र वाप्याम्॥ २२२॥

न्यासः 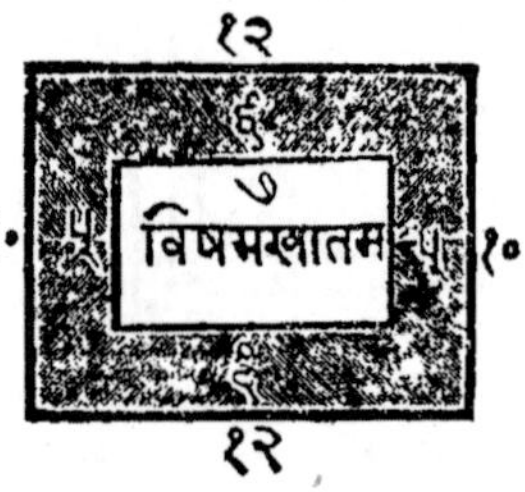

मुखजं क्षेत्रफलम्। १२०। तलजम्। ३०। तद्युतिजम्। २७०। एषामैक्यम्। ४२०।

षड्भिः। ६। हृतं जातं समफलम्। ७०। वेध। ७। हतं जातं खातफलं घनहस्ताः। ४९०।

द्वितीयोदाहरणम्। खातेऽथ तिग्मकरतुल्यचतुर्भुजे च किं स्यात् फलं नवमितः किल यत्र वेधः। वृत्ते तथैव दशविस्तृतिपञ्चवेधे सूचीफलं वद तयोश्च पृथक् पृथङ्मे॥ २२३॥

न्यासः 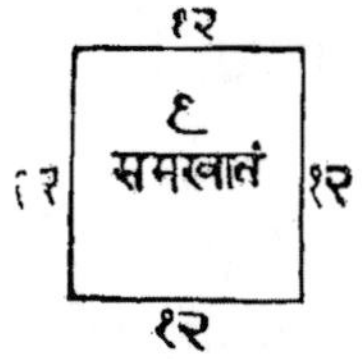

भुजः। १२। वेधः। ९।
जातं यथोक्तकरणेन खात-
फलं घनहस्ताः। १२९६।
सूचीफलम्। ४३२।

वृत्तखातदर्शनाय

न्यासः 

व्यासः। १०। वेधः। ५।
अत्र सूक्ष्मपरिधिः
$\frac{३९२७}{१२५}$

सूक्ष्मक्षेत्रफलम्
$\frac{३९२७}{५०}$

वेधगुणं जातं सूक्ष्मखातफलम् $\frac{३९२७}{१०}$ सूक्ष्मसूचीफलम्
$\frac{१३०९}{१०}$

यद्वा स्थूलखातफलम् $\frac{२७५०}{७}$ सूचीफलं स्थूलं वा $\frac{२७५०}{२१}$

इति खातव्यवहारः।

[अष्टमोऽध्यायः।]

चितौ करणसूत्रं सार्द्धवृत्तम्। उच्छ्रयेण गुणितं चितेः किल क्षेत्रसम्भवफलं घनं भवेत्। इष्टिकाघनहृते घने चितेरिष्टिकापरिमितिश्च लभ्यते। इष्टिकोच्छ्रयहृदुच्छ्रितिश्चितेः स्युस्तराश्च दृषदां चितेरपि॥ २२४-२२५॥

उदाहरणम्। अष्टादशाङ्गुलं दैर्घ्यं विस्तारो द्वादशाङ्गुलः। उच्छ्रितिस्त्र्यङ्गुला यासामिष्टिकास्ताश्चितौ किल॥ यद्विस्तृतिः पञ्चकराष्टहस्तं दैर्घ्यं च यस्यां त्रिकरोच्छ्रितिश्च। तस्यां चितौ किं फलमिष्टिकानां सङ्ख्या च का ब्रूहि कतिस्तराश्च॥ २२६-२२७॥

न्यासः

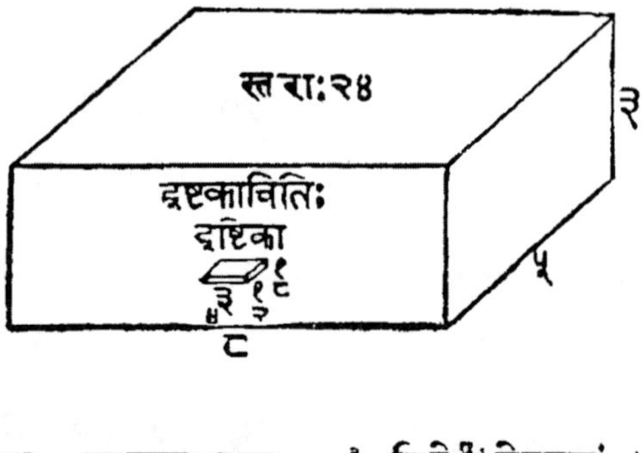

इष्टिकाया घनहस्तमानम् ३/६४ चितेः [illegible] क्षेत्रफलं १। ४०। उच्छ्रयेण गुणितं चितेर्घनफलम्। १२०। लब्धा इष्टिकासङ्ख्या। २५६०। स्तरसङ्ख्या। २४। एवं पाषाणचयेऽपि।

इति चितिव्यवहारः।

[नवमोऽध्यायः।]

अथ क्राकचव्यवहारे करणसूत्रं वृत्तम्। पिण्डयोगदलमग्रमूलयोः दैर्घ्यसङ्गुणितमङ्गुलात्मकम्। दारुदारणपथैः समाहतं षट्स्वरेषु विहृतं करात्मकम्॥ २२८॥

उदाहरणम्। मूले नखाङ्गुलमितोऽथ नृपाङ्गुलोऽग्रे पिण्डः शताङ्गुलमितं किल यस्य दैर्घ्यम्। तद्दारुदारणपथेषु चतुर्षु किं स्याद्धस्तात्मकं वद सखे गणितं द्रुतं मे॥ २२९॥

न्यासः

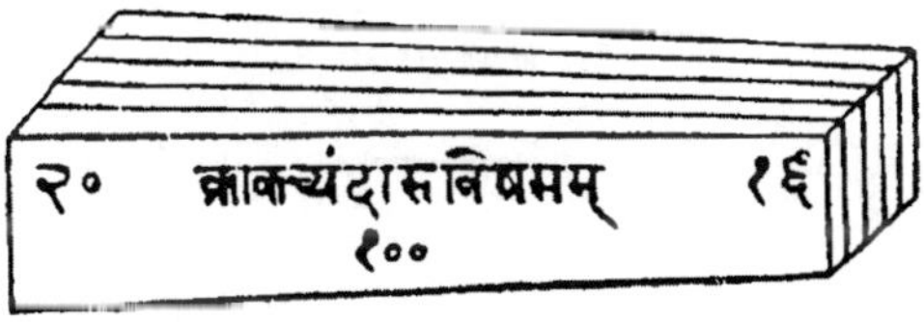

पिण्डयोगदलं। १८। दैर्घ्येण। १००। सङ्गुणितं। १८००। दारुदारणपथैः। ४। गुणितम्। ७२००। षट्स्वरेषु। ५७६। विहृतं जातं करात्मकं गणितम् $\frac{३५}{२}$

क्राकचान्तरे करणसूत्रं सार्धवृत्तम्। छिद्यते तु यदि तिर्यगुक्तवत्पिण्डविस्तृतिहतेः फलं तदा॥ २३०॥

इष्टिकाचितिदृषच्चितिखातक्राकचव्यवहृतौ खलु मूल्यम्। कर्मकारजनसम्प्रतिपत्त्या तन्मृदुत्वकठिनत्ववशेन॥ २३१॥

उदाहरणम्। यद्विस्तृतिर्दन्तमिताङ्गुलानि पिण्डस्तथा षोड़श यत्र काष्ठे। छेदेषु तिर्य्यङ्नवसु प्रचक्ष्व किं स्यात् फलं तत्र करात्मकं मे॥ २३२॥

न्यासः 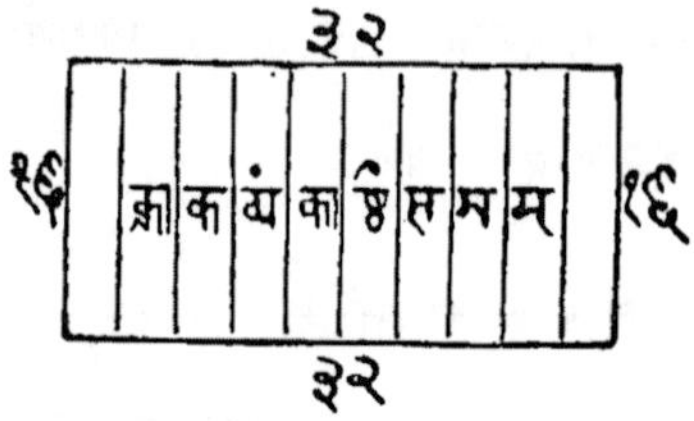

विस्तारः। ३२। पिण्डः। १६।

पिण्डविस्तृतिहतिः। ५१२। दारुदारणमार्ग-। ९। घा। ४६०८। षट्खरेषु। ५७६। विह्रतं जातं फलं हस्ताः। ८।

इति क्रकचव्यवहारः।

[दशमोऽध्यायः।]

अथ राशिव्यवहारे करणसूत्रं वृत्तम्। अनणुषु दशमांशोऽणुष्वथैकादशांशः परिधिनवमभागः शूकधान्येषु वेधः। भवति परिधिषष्ठे वर्गिते वेधनिघ्ने घनगणितकराः। स्युर्मागधास्ताश्च खार्य्यः॥ २३२॥

उदाहरणम्। समभुवि किल राशिर्यः स्थितः स्थूलधान्यः परिधिपरिमितिः स्याद्धस्तषष्टिर्यदीया। प्रवद गणक खार्य्यः किम्मिताः सन्ति तस्मिन्नथ पृथगणुधान्यैः शूकधान्यैश्च शीघ्रम्॥ २३४॥

अथ स्थूलधान्यराशिमानावबोधनाय

न्यासः

परिधिः। ६०। वेधः। ६।
परिधेः षष्ठांशः। १०।
वर्गितः। १००। वेधनिघ्नः
लब्धाः खार्य्यः। ६००।

अथाणुधान्यराशिमानानयनाय

न्यासः

परिधिः। ६०। वेधः $\frac{६०}{११}$
जातं फलम् ५४५ $\frac{५}{११}$

अथ शूकधान्यराशिमानानयनाय

न्यासः

परिधिः। ६०।
वेधः $\frac{२०}{३}$ खार्य्यः

६६६ $\frac{२}{३}$

अथ भित्त्यन्तर्बाह्यकोणसंलग्नराशिप्रमाणानयनकरणसूत्रं वृत्तम्।

द्विवेदसत्रिभागैकनिघ्नात् परिधेः फलम् । भित्त्यन्तर्बाह्यक राशेः स्वगुणभाजितम् ॥ २३५ ॥

उदाहरणम् । परिधिर्भित्तिलग्नस्य राशेस्त्रिंशत्करः किल। अन्तःकोणस्थितस्यापि तिथितुल्यकरः सखे॥ बहिःकोणस्थितस्यापि पञ्चघ्ननवसम्मितः। तेषामाचक्ष्व मे क्षिप्रं घनहस्तान् पृथक् पृथक् ॥ २३६-२३७ ॥

अत्रापि स्थूलादिधान्यानां राशिमानावबोधनाय स्पष्टं चेत्रत्रयम्।

तत्रादावनणुधान्यराशिमानबोधकं चेत्रम्।

न्यासः

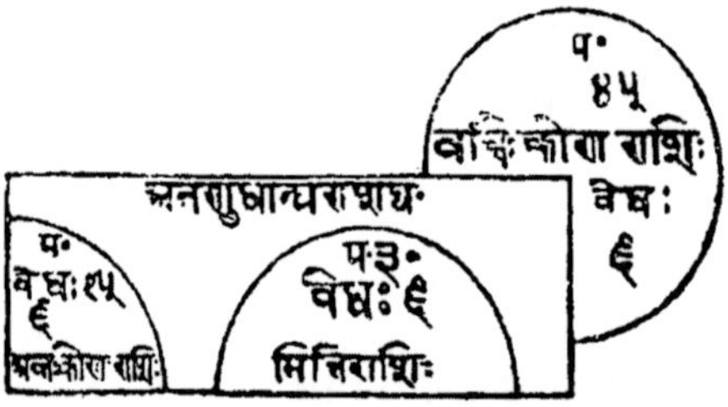

अत्राद्यस्य परिधिः। ३०। द्विनिघ्नः। ६०। अन्य-। १५। चतुर्घ्नः। ६०। अपरः। ४५। सत्रिभागैक- $\frac{४}{३}$ निघ्नः। ६०। एषां वेधः। ६। एभ्यः फलं तुल्यमेतावन्त्य एव खार्य्यः। ६००। एतत् स्वस्वगुणेन भक्तं जातं पृथक् पृथक् फलम्। ३००। १५०। ४५०।

अथाणुधान्यराशिमानानयनाय

न्यासः

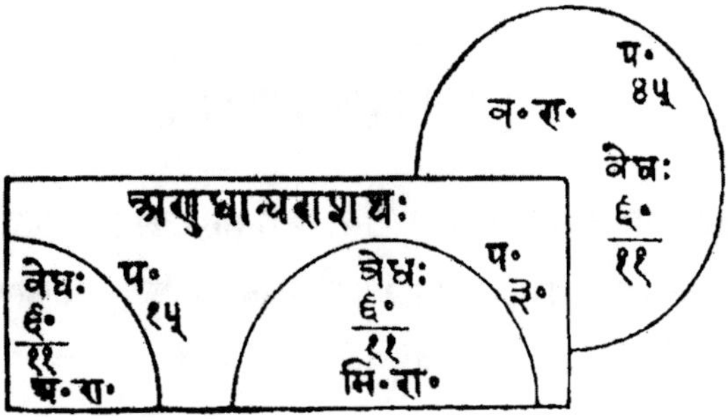

पूर्ववत् क्षेत्रत्रयाणां स्वगुणगुणितपरिधिः। ६०। वेधः $\frac{६०}{११}$ फलानि २७२ $\frac{८}{११}$ १३६ $\frac{४}{११}$ ४०९ $\frac{१}{११}$

अथ शूकधान्यराशिमानानयनाय

न्यासः

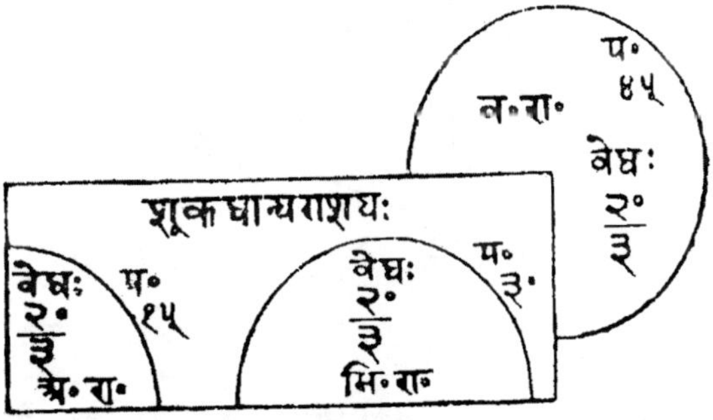

अत्रापि पूर्ववत् क्षेत्रत्रयाणां स्वगुणगुणितः परिधिः। ६०। वेधः $\frac{२०}{३}$ फलानि ३३३ $\frac{१}{३}$ १६६ $\frac{२}{३}$। ५००।

इति राशिव्यवहारः समाप्तः।

[एकादशोऽध्यायः।]

अथ छायाव्यवहारे करणसूत्रं वृत्तम्। छाययोः कर्णयोरन्तरे ये तयोर्वर्गविश्लेषभक्ता रसाद्रीषवः। सैकलब्धेः पदघ्नन्तु कर्णान्तरं भान्तरेणोनयुक्तं दले स्तः प्रभे ॥ २३८ ॥

उदाहरणम्। नन्दचन्द्रैर्मितं छाययोरन्तरं कर्णयोरन्तरं विश्वतुल्यं ययोः। ते प्रभे वक्ति योर्युक्तिमान् वेत्त्यसौ व्यक्तमव्यक्तयुक्तं हि मन्येऽखिलम् ॥ २३९ ॥

न्यासः 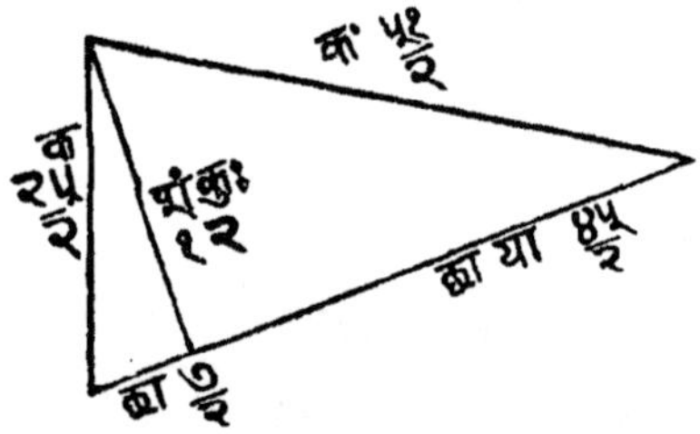

छायान्तरम्। १९। कर्णान्तरम्। १३। अनयोर्वर्गान्तरेण। १९२। भक्ता रसाद्रीषवः। ५७६। लब्धम्। ३। सैकस्यास्य। ४। मूलम्। २। अनेन कर्णान्तरं। १३। गुणितम्। २६। द्विस्थम्। २६। भान्तरेण। १९। ऊनयुते। ७। ४५। तदर्धे लब्धे छाये $\frac{७}{२}$ $\frac{४५}{२}$ तत्कृत्योर्योगपदमित्यादिना जातौ कर्णौ $\frac{२५}{२}$ $\frac{५१}{२}$

छायान्तरे करणसूत्रं वृत्तार्द्धम्। शङ्कुः प्रदीपतलशङ्कुतलान्तरघ्न-श्छाया भवेद्विनरदीपशिखौच्च्यभक्तः ॥ २४० ॥

उदाहरणम्। शङ्कुप्रदीपान्तरभूस्त्रिहस्ता दीपोच्छ्रितिः सार्द्ध-करत्रया चेत्। शङ्कोस्तदार्काङ्गुलसम्मितस्य तस्य प्रभा स्यात्कियती वदाशु ॥ २४१ ॥

न्यासः

शङ्कुः $\frac{१}{२}$ प्रदीपशङ्कुतलान्तरम्। ३। अनयोर्घातः $\frac{३}{२}$ विनर-दीपशिखौच्च्येन। ३। भक्तो लब्धानि छायाङ्गुलानि। १२।

अथ दीपोच्छ्रित्यानयनाय करणसूत्रं वृत्तार्द्धम्। छायाहृते तु नरदीपतलान्तरघ्ने शङ्कौ भवेन्नरयुते खलु दीपकौच्च्यम् ॥ २४२ ॥

उदाहरणम्। प्रदीपशङ्क्वन्तरभुस्त्रिहस्ता छायाङ्गुलैः षोड़शभिः समा चेत्। दीपोच्छ्रितिः स्यात् कियती वदाशु प्रदीपशङ्क्वन्तर-मुच्यतां मे ॥ २४३ ॥

न्यासः

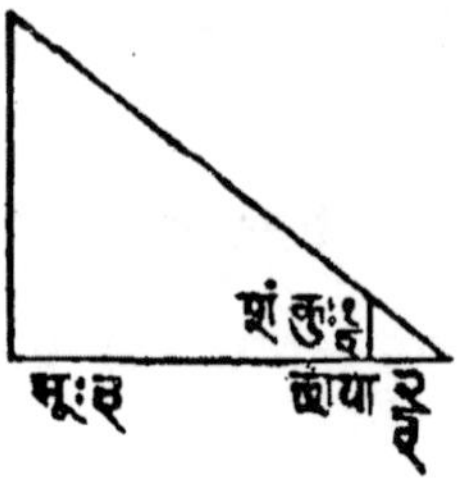

शङ्कुः। १२। छायाङ्गुलानि। १६। शङ्कुप्रदीपान्तरहस्ताः। लब्धं दीपकौच्च्य हस्ताः ११/४

प्रदीपशङ्क्वन्तरभूमानानयनाय करणसूत्रं वृत्तार्द्धम्। विशङ्कुदीपोच्छ्रयसङ्गुणा भा शङ्कूद्धृता दीपनरान्तरं स्यात्॥ २४४॥

उदाहरणम्। पूर्वोक्त एव दीपोच्छ्रायः ११/४. शङ्क्वङ्गुलानि। १२। छाया। १६। लब्धाः शङ्कुप्रदीपान्तरहस्ताः। ३।

छायाप्रदीपान्तरदीपौच्च्यानयनाय करणसूत्रं सार्द्धवृत्तम्। छायाग्रयोरन्तरसङ्गुणाभा छायाप्रमाणान्तरहृद्भवेद्भूः। भूशङ्कुघातः प्रभया विभक्तः प्रजायते दीपशिखौच्च्यमेवम्। त्रैराशिकेनैव यदेतदुक्तं व्याप्तं स्वभेदैर्हरिणेव विश्वम्॥ २४५॥

उदाहरणम्। शङ्कोर्भार्कमिताङ्गुलस्य सुमते दृष्टा किलाष्टाङ्गुला छायाग्राभिमुखे करद्वयमिते न्यस्तस्य देशे पुनः। तस्यैवार्कमिता-

ङ्गुला यदि तदा छायाप्रदीपान्तरं दीपौच्चयञ्च कियद्वद व्यवहृतिं छायाभिधां वेत्सि चेत्॥ २४६॥

न्यासः

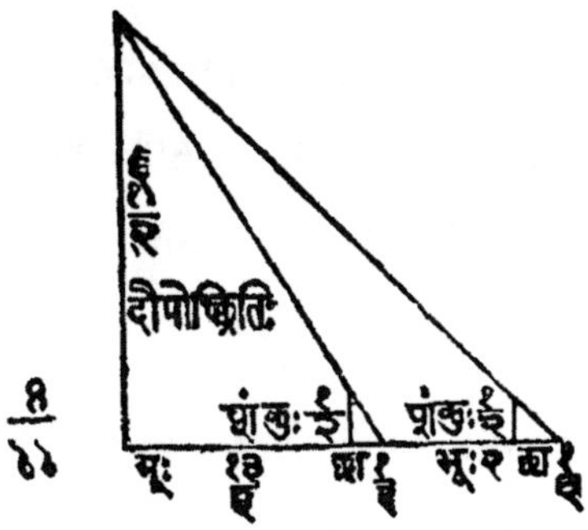

अत्र छायाग्रयोरन्तरमङ्गुलात्मकम्। ५२। छाये च। ८। १२। अनयोराद्या। ८। इयमनेन। ५२। गुणिता। ४१६। छायाप्रमाणान्तरेण। ४। भक्ताल्लब्धं भूमानम्। १०४। इदं प्रथमच्छायाग्रदीपतलयोरन्तरमित्यर्थः। एवं द्वितीयच्छायाग्रमानम्। १५६। भूशङ्कुघातः प्रभया विभक्तइति जातमुभयतोऽपि दीपौच्चंग समम

हस्ताः ६ $\frac{1}{2}$

एवमित्यत्र छायाव्यवहारे त्रैराशिककल्पनयानयनं वर्त्तते तद्यथा। प्रथमच्छायातोद्वितीयच्छाया यावताधिका तावता छायावयवेन यदि छायाग्रान्तरतुल्याभूर्लभ्यते तदा प्रथमच्छायया किमिति एवं

पृथक् पृथक् छायाप्रदीपतलान्तरप्रमाणं लभ्यते। ततो द्वितीयं त्रैराशिकम्। यदि छायातुल्ये भुजे शङ्कुः कोटिस्तदा भूतुल्ये भुजे किमिति लब्धं दीपकौच्च्यमुभयतोऽपि तुल्यमेव। एवं पञ्चराशिकादि-कर्माखिलं त्रैराशिककल्पनयैव सिद्धम्। यथा भगवता श्रीनारायणेन जननमरणक्लेशापहारिणा निखिलजगज्जननैकबीजेन सकलभुवन-भावनेन गिरिसरित्सुरनरासुरादिभिः स्वभेदैरिदं जगद्व्याप्तं तथेदमखिलं गणितजातं त्रैराशिकेन व्याप्तं यद्येवं तद्बहुभिः किमित्याशङ्क्याह।

यत्किञ्चिद्गुणभागहारविधिना बीजेऽत्र वा गण्यते तत्त्रैराशिकमेव निर्मलधियामेवावगम्यं विदाम्। एतद् यद्बहुधास्मदादिजडधीधी-वृद्धिबुद्ध्या बुधैस्तद्भेदान् सुगमान्विधाय रचितं प्राज्ञैः प्रकीर्णादि-कम्॥ २४७॥

इति छायाधिकारः।

[द्वादशोऽध्यायः।]

अथ कुट्टके करणसूत्रं वृत्तपञ्चकम्। भाज्यो हारः क्षेपकश्चापवर्त्यः केनाप्यादौ सम्भवे कुट्टकार्थम्। येन च्छिन्नौ भाज्यहारौ न तेन क्षेपश्चैतद्दुष्टमुद्दिष्टमेव॥ २४८॥

---

१ 'सकलभुवनभवन' इति पाठान्तरम्।

परस्परं भाजितयोर्ययोर्यः शेषस्तयोः स्यादपवर्त्तनं सः। तेनापवर्त्तेन विभाजितौ यौ तौ भाज्यहारौ दृढसंज्ञकौ स्तः॥ २४९॥

मिथोभजेतौ दृढभाज्यहारौ यावद्विभाज्ये भवतीह रूपम्। फलान्यधोऽधस्तदधोनिवेश्यः क्षेपस्ततः शून्यमुपान्तिमेन॥ २५०॥

स्वोर्द्ध्वे हतेऽन्त्येन युते तदन्त्यं त्यजेन्मुहुः स्यादिति राशियुग्मम्। ऊर्ध्वो विभाज्येन दृढेन तष्टः फलं गुणः स्यादधरो हरेण॥ २५१॥

एवं तदैवात्र यदा समास्ताः स्युर्लब्धयश्चेद्विषमास्तदानीम्। यदागतौ लब्धिगुणौ विशोध्यौ स्वतक्षणाच्छेषमितौ तु तौ स्तः॥ २५२॥

उदाहरणम्। एकविंशतियुतं शतद्वयं यद्गुणं गणक पञ्चषष्टियुक्। पञ्चवर्जितशतद्वयोद्धृतं शुद्धिमेति गुणकं वदाशु तम्॥ २५२॥

न्यासः। भाज्यः। २२१। हारः। १९५। क्षेपः। ६५। अत्र परस्परभाजितयोर्भाज्यभाजकयोः शेषम्। १३। अनेन भाज्यहारक्षेपा अपवर्त्तिता जाता भाज्यः। १७। हारः। क्षेपः। ५। अनयोः दृढभाज्यहारयोः परस्परभक्तयोर्लब्ध्यान्यधोऽधस्तदधः क्षेपस्तदधः शून्यं

निवेश्यमिति न्यस्ते जाता वल्ली १ उपान्तिमेन स्वोर्ध्वे
७
५
०

हते इत्यादिकरणेन जातं राशिद्वयम् ४० एतौ दृढभाज्य-
३५

हाराभ्यां १७ तष्टौ लब्धिगुणौ जातौ। ६। ५। इष्टाहत-
१५

स्वस्वहरेण युक्ते इति वक्ष्यमाणविधिनैताविष्टगुणितस्वतक्षणयुक्तौ वा लब्धिगुणौ। २३। २०। द्विकेनेष्टेन वा। ४०। ३५। इत्यादि।

कुट्टकान्तरे करणसूत्रं वृत्तम्। भवति कुट्टविधेर्युतिभाज्ययोः समपवर्त्तितयोरपि वा गुणः। भवति यो युतिभाजकयोः पुनः स च भवेदपवर्त्तनसङ्गुणः॥ २५४॥

उदाहरणम्। शतं हतं येन युतं नवत्या विवर्जितं वा विहृतं त्रिषष्ट्या। निरग्रकं स्याद्वद मे गुणं तं स्पष्टं पटीयान् यदि कुट्टकेऽसि॥ २५५॥

न्यासः। भाज्यः। १००। हारः। ६३। क्षेपः। ९०। जाता पूर्ववल्लब्धिक्षेपाणां

वल्ली १
१
१
२
२
१
९०
०

उपान्तिमेन स्वोर्द्धे हतेऽन्त्येन युत इत्यादिकरणेन जातं राशिद्वयम्
२४३० जातौ पूर्ववल्लब्धिगुणौ। ३०। १८।
२५३०

अथवा भाज्यक्षेपौ दशभिरपवर्त्तय भाज्यः। १०। क्षेपः।
८। परस्परभजनाल्लब्धानि फलानि क्षेपं शून्यञ्चाधोऽधो निवेश्य
जाता
वल्ली ०
६
३
८
०

पूर्ववल्लब्धो गुणः। ४५। अत्र लब्धिर्न ग्राह्या। यतो लब्धयो
विषमा जाता अतो गुणे। ४५। स्वतक्षणादस्माद्। ६३।
विशोधिते जातो गुणः स एव। १८। गुणघ्नभाज्ये क्षेपयुते
हरतष्टे लब्धिश्च। ३०।

अथवा हारक्षेपौ। ६३। ९०। नवभिरपवर्त्तितौ जातौ
हारक्षेपौ। ७। १०। अत्र लब्धिक्षेपाणां
वल्ली १४
३
१०
०

लब्धो गुणः। २। क्षेपहारापवर्त्तनेन। ८। गुणितो जातः
१४

स एव गुणः। १८। भाज्य-। १००। भाजक-। ६३। क्षेपेभ्यो। ९०। लब्धिश्च। ३०।

अथवा भाज्यक्षेपौ पुनर्हारक्षेपौ चापवर्त्तितौ जातौ भाज्यहारौ। १०। ७। क्षेपः। १। अत्र पूर्ववज्जाता

वल्ली १
२
१
०

गुणश्च। २। हारक्षेपापवर्त्तनेन गुणितो जातः स एव गुणः। १८। पूर्ववल्लब्धिश्च। ३०। इष्टाहतस्वस्वहरेण युक्ते इत्यादिनाथवा गुणलब्धी। ८१। १३०। इत्यादि।

कुट्टकान्तरे करणसूत्रं वृत्तार्द्धम्। क्षेपजे तक्षणाच्छुद्धे गुणाप्ती स्तो वियोगजे॥ २५६॥

अत्र पूर्वोदाहरणे नवतिक्षेपे यौ लब्धिगुणौ जातौ। ३०। १८। एतौ स्वतक्षणाभ्यामाभ्यां। १००। ६३। शोधितौ मे शेषके तन्मितौ लब्धिगुणौ नवतिशोधने ज्ञातव्यौ। ७०। ४५। एतयोरपि इष्टाहतस्वस्वतक्षणं क्षेप इति लब्धिगुणौ। १७०। १०८। अथवा। २७०। १७१। इत्यादि।

द्वितीयोदाहरणम्। यद्गुणा गणक षष्टिरन्विता वर्जिता च दशभिः षडुत्तरैः। स्यात्त्रयोदशहृता निरग्रका तद्गुणं कथय मे पृथक् पृथक्॥ २५७॥

न्यासः। भाज्यः। ६०। हारः। १३। क्षेपः। १६।
प्राग्वज्जाता वल्ली ४
१
१
१
१
१६
०

तथा जाते गुणाप्ती। २। ८। अत्रापि लब्ध्योर्विषमाः। अतो गुणाप्ती स्वतक्षणाभ्यां। १३। ६०। शोधिते जाते। ११। ५२। एवं षोडशक्षेपे एतौ एव लब्धिगुणौ। ५२। ११। स्वस्वहराभ्यां शोधितौ जातौ षोडशविशुद्धौ। २। ८।

कुट्टकान्तरे करणसूत्रं सार्द्धवृत्तम्। गुणलब्ध्योः समं ग्राह्यं धीमता तक्षणे फलम्। हरतष्टे धनक्षेपे गुणलब्धी तु पूर्ववत्। क्षेपतक्षणलाभाढ्या लब्धिः शुद्धौ तु वर्जिता॥ २५८॥

उदाहरणम्। येन सङ्गुणिताः पञ्च त्रयोविंशतिसंयुताः। वर्जिता वा त्रिभिर्भक्ता निरग्राः स्युः स को गुणः॥ २५९॥

न्यासः। भाज्यः। ५। हारः। ३। क्षेपः। २३। अत्र वल्ली १
१
२३
०

पूर्ववज्जातं राशिद्वयम् $\frac{४६}{२३}$ एतौ भाज्यहाराभ्यां तष्टौ।

अत्राधोराशौ। २३। त्रिभिस्तष्टे सप्त लभ्यन्ते ऊर्ध्वराशौ। ४६।
पञ्चभिस्तष्टे नव लभ्यन्ते तत्र नव न ग्राह्याः गुणलब्ध्योः समं ग्राह्यं
धीमता तक्षणे फलमिति अतः सप्तैव ग्राह्याः। एवं जातौ गुणाप्ती।
२। ११। क्षेपजे तक्षणाच्छुद्धे इति त्रयोविंशतिशुद्धौ जाता विपरीत-
शोधनादवशिष्टा लब्धिः। ६। शुद्धौ जाते। १। ६।

इष्टाहतस्वस्वहरेण युक्ते इति वक्ष्यमाणविधिना धनर्णयोरन्तरमेव
योगः इति बीजोक्त्या च इष्टगुणितस्वहारक्षेपणेन यथा धनलब्धिः
स्यादिति तथा कृते जातौ गुणाप्ती। ७। ४। एवं सर्वत्र।

अथवा हरतष्टे धनक्षेपे इति

न्यासः। भाज्यः। ५। हारः। ३। क्षेपः। २।

पूर्ववज्जाते गुणाप्ती। २। ४। एते स्वस्वहाराभ्यां शोधिते
विशुद्धिजे जाते। १। १। क्षेपतक्षणलाभाढ्या लब्धिरिति जातौ
क्षेपजौ लब्धिगुणौ। ११। २। शुद्धौ तु वर्जितेति शुद्धिजौ भवतः
किन्त्वत्र शुद्धा न भवति तस्माद्विपरीतशोधनेन ऋणलब्धिः। ६।
गुणः। १। धनलब्ध्यर्थं द्विगुणं स्वहारे क्षिप्ते सति जाते।
। ७। ४।

कुट्टकान्तरे करणसूत्रं वृत्तम्। क्षेपाभावोऽथवा यत्र क्षेपः शुद्धोहरोद्धृतः। ज्ञेयः शून्यं गुणस्तत्र क्षेपो हारहृतः फलम्॥ २६०॥

उदाहरणम्। येन पञ्च गुणिताः खसंयुताः पञ्चषष्टिसहिताश्च तेऽथवा। स्युस्त्रयोदशहृतानिरग्रकास्तं गुणं गणक कीर्त्तयाशु मे॥ २६१॥

न्यासः। भाज्यः। ५। हारः। १३। क्षेपः। ०। ज्ञेयः शून्यं गुणस्तत्र क्षेपोहारहृतः फलमिति क्षेपाभावे गुणाप्ती। ०। ०। इष्टाहतेत्यथवा। १३। ५। वा। २६। १०।

न्यासः। भाज्यः। ५। हारः। १३। क्षेपः। ६५।

क्षेपः शुद्धोहरोद्धृतः ज्ञेयः शून्यं गुणस्तत्र क्षेपोहारहृतः फलमिति जाते गुणाप्ती। ०। ५। वा। १३। १०। अथवा। २६। २५। इत्यादि।

अथ सर्वत्र कुट्टके गुणलब्ध्योरनेकधा दर्शनार्थं करणसूत्रं वृत्तार्द्धम्। इष्टहतस्वस्वहरेण युक्ते ते वा भवेतां बहुधा गुणाप्ती॥ २६२॥

अस्योदाहरणानि दर्शितानि पूर्वमिति।

अथ स्थिरकुट्टके करणसूत्रं वृत्तम्। क्षेपे तु रूपे यदि वा विशुद्धे स्यातां क्रमाद् ये गुणकारलब्धी। अभीप्सितक्षेपविशुद्धिनिघ्ने स्वहारतष्टे भवतस्तयोस्ते॥ २६२॥

प्रथमोदाहरणे दृढ़भाज्याहारयोः रूपक्षेपयोर्न्यासः भाज्यः। १७। हारः। १५। क्षेपः। १।

अत्र गुणाप्तौ। ७। ८। एते त्विष्टक्षेपेण पञ्चकेन गुणिते स्वहारतष्टे च जाते। ५। ६। अथ रूपशुद्धौ गुणाप्ती। ८। ७। तक्षणाच्छुद्धौ जातौ लब्धिगुणौ। ९। ८। एते पञ्चगुणे स्वहारतष्टे च जाते। १०। ११। एवं सर्वत्र।

अस्य ग्रहगणिते उपयोगस्तदर्थं किञ्चिदुच्यते। कल्प्याथ शुद्धिर्विकलावशेषः षष्टिश्च भाज्यः कुदिनानि हारः। तज्जं फलं स्युर्विकलागुणस्तु लिप्ताग्रमस्माच्च कलालवाग्रम्। एवं तदूर्ध्वञ्च तथाधिमासावमाग्रकाभ्यां दिवसा रवीन्द्वोः॥ २६४॥

ग्रहस्य विकलावशेषात् ग्रहाहर्गणयोरानयनं तद्यथा। तत्र षष्टिर्भाज्यः कुदिनानि हारः विकलावशेषं शुद्धिरिति प्रकल्प्य साध्ये गुणाप्ती। तत्र लब्धिर्विकलाः स्युर्गुणस्तु कलावशेषम्।

एवं कलावशेषं शुद्धिस्तत्र षष्टिर्भाज्यः कुदिनानि हारः लब्धिः कला गुणो भागशेषम्।

भागशेषं शुद्धिः त्रिंशद्भाज्यः कुदिनानि हारः फलं भागाः गुणो-राशिशेषम्।

एवं राशिशेषं शुद्धिः द्वादश भाज्यः कुदिनानि हारः फलं गतराशयः गुणो भगणशेषम्।

कल्पभगणोभाज्यः कुदिनानि हारः भगणशेषं शुद्धिः कलं गतभगणाः गुणोऽहर्गणः स्यादिति।

अस्योदाहरणानि त्रिप्रश्नाध्याये।

एवं कल्पाधिमासाः भाज्यः रविदिनानि हारः अधिमासशेषं शुद्धिः फलं गताधिमासाः गुणो मतरविदिवसाः।

एवं कल्पावमानि भाज्यः चन्द्रदिवसाहारः अवमशेषं शुद्धिः फलं गतावमानि गुणोगतचान्द्रदिवसा इति।

संश्लिष्टकुट्टके करणसूत्रं वृत्तम्। एको हरश्चेद् गुणकौ विभिन्नौ तदा गुणैक्यं परिकल्प्य भाज्यम्। अग्रैक्यमग्रं कृम उक्तवद् यः संश्लिष्टसंज्ञः स्फुटकुट्टकोऽसौ॥ २६५॥

उदाहरणम्। कः पञ्चनिघ्नो विहृत त्रिषष्ट्या सप्तावशेषोऽथ स एव राशिः। दशाहतः स्याद्विहृतस्त्रिषष्ट्या चतुर्दशाग्रो वद राशिमेनम्॥ २६६॥

अत्र गुणैक्यं भाज्यः अग्रैक्यं शुद्धिरिति।

न्यासः। भाज्यः। १५। हारः। ६३। शुद्धिः। २१। पूर्ववज्जातः शुद्धौ गुणः। १४।

इति लीलावत्यां कुट्टकाध्यायः।

[त्रयोदशोऽध्यायः ।]

अथ गणितपाशे निर्दिष्टाङ्कैः सङ्ख्याया विभेदे करणसूत्रं वृत्तम्।

स्थानान्तमेकादिचयाङ्कघातः सङ्ख्याविभेदानियतैः स्युरङ्कैः।
भक्तोऽङ्कमित्याङ्कसमासनिघ्नः स्थानेषु युक्तोमितिसंयुतिः स्यात्॥ २६७॥

अत्रोद्देशकः। द्विकाष्टकाभ्यां त्रिनवाष्टकैर्वा निरन्तरं द्व्यादिन-वावसानैः। सङ्ख्याविभेदाः कति सम्भवन्ति तत्सङ्ख्यकैक्यानि पृथग्वदाशु॥ २६८॥

न्यासः। २। ८। अत्र स्थाने। स्थानान्तमेकादिचयाङ्कयोः। ।१।२। घातः।२। एवं जातौ सङ्ख्याभेदौ।२। अथ स एव घातोऽङ्कसमास-।१०। निघ्नः।२०। अङ्कमित्यानया २। भक्तः।१०। स्थानद्वयेयुक्तोजातं सङ्ख्यैक्यम्।११०।

द्वितीयोदाहरणे न्यासः।३।९।८। अत्रैकादिचयाङ्कानाम्। १।२।३। घातः।६। एतावन्तः सङ्ख्याभेदाः। अथ स एव घातः।६। अङ्कसमासा-।२०। हतः।१२०। अङ्कमित्या।३। भक्तः।४०। स्थानत्रये युक्तोजातं सङ्ख्यैक्यम्।४४४०।

तृतीयोदाहरणे न्यासः।२।३।४।५।६।७।८।९। एवमत्र सङ्ख्याभेदाश्चत्वारिंशत्सहस्राणि शतत्रयं विंशतिश्च।४०३२०।

सङ्ख्यैक्यञ्च चतुर्विंशति निखर्वाणि त्रिषष्टिपद्मानि नवनवतिकोटयः नवनवतिलक्षाणि पञ्चसप्ततिसहस्राणि शतत्रयं षष्टिश्च। २४६३-९९९९७५३६०।

उदाहरणम्। पाशाङ्कुशाहिडमरूककपालशूलैः खट्वाङ्गशक्ति-शरचापयुतैर्भवन्ति। अन्योन्यहस्तकलितैः कतिमूर्त्तिभेदाः शम्भो-र्हरेरिव गदारिसरोजशङ्खैः॥ २६९॥

न्यासः। स्थानानि। १०। जातामूर्त्तिभेदाः। ३६२८८००। एवं हरेश्च। २४।

विशेषकरणसूत्रं वृत्तम्। यावत्स्थानेषु तुल्याङ्कास्तद्भेदैस्तु पृथक् कृतैः। प्राग्भेदाविहृताभेदास्तत्सङ्ख्यैक्यञ्च पूर्ववत्॥ २७०॥

अत्रोद्देशकः। द्विद्व्येकभूपरिमितैः कतिसङ्ख्यकाः स्युस्तासां युतिञ्च गणकाशु मम प्रचक्ष्व। अम्भोधिकुम्भिशरभूतशरैस्तथाङ्कैश्चेद-ङ्कपाशमितियुक्तिविशारदोऽसि॥ २७१॥

न्यासः। २। २। १। १। अत्र प्राग्वद्भेदाः। २४। यावत् स्थानेषु तुल्याङ्का इत्यत्र प्रथमं तावत् स्थानद्वये तुल्यौ प्राग्वत् स्थान द्वयाज्जातौ भेदौ। २। पुनरत्रापि स्थानद्वये तुल्यौ तत्राप्येवं भेदौ।

२। भेदाभ्यां प्रागभेदाः। २४। भक्ता जाता भेदाः। ६। तद्यथा। २२११। २१२१। २११२। १२१२। ११२२। पूर्ववत् सङ्ख्यैक्यञ्च। ९९९९।

द्वितीयोदाहरणे न्यासः। ४। ८। ५। ५। ५। अत्रापि पूर्ववद्भेदाः। १२०। स्थानत्रयोत्थभेदैः। ६। भक्ता जाताः। २०। तद्यथा

। ४ ८ ५ ५ ५ । ८ ४ ५ ५ ५ । ५ ४ ८ ५ ५ ।
। ५ ८ ४ ५ ५ । ५ ५ ४ ८ ५ । ५ ५ ८ ४ ५ ।
। ५ ५ ५ ४ ८ । ५ ५ ५ ८ ४ । ४ ५ ८ ५ ५ ।
। ४ ५ ५ ८ ५ । ४ ५ ५ ५ ८ । ८ ५ ४ ५ ५ ।
। ८ ५ ५ ४ ५ । ८ ५ ५ ५ ४ । ५ ४ ५ ८ ५ ।
। ५ ८ ५ ४ ५ । ५ ५ ४ ५ ८ । ५ ५ ८ ५ ४ ।
। ५ ४ ५ ५ ८ । ५ ८ ५ ५ ४ । एवं विंशति।

अथ सङ्ख्यैक्यञ्च। ११९९९८८।

अनियताङ्कैरतुल्यैश्च विभेदे करणसूत्रं वृत्तार्धम्। स्थानान्त-मेकापचितान्तिमाङ्कघातोऽसमाङ्कैश्च मितिप्रभेदाः॥ २७२॥

उदाहरणम्। स्थानषट्कस्थितैरङ्कैरन्योन्यं खेन वर्जितैः। कति सङ्ख्याविभेदाः स्युर्यदि वेत्सि निगद्यताम्॥ २७३॥

अत्रान्तिमाङ्को नव षट्स्थाने एकापचितान्यस्ता न्यासः। ९। ८। ७। ६। ५। ४। एषां घाते जाताः सङ्ख्याभेदाः। ६०४८०।

अन्यत् करणसूत्रं वृत्तद्वयम्। निरेकमङ्कैक्यमिदं निरेकस्थानान्तमेकापचितं विभक्तम्। रूपादिभिस्तन्निहतैः समाः स्युःसङ्ख्याविभेदा नियतेऽङ्कयोगे॥ नवान्वितस्थानकसङ्ख्यकाया ऊनेऽङ्कयोगे कथितं तु वेद्यम्। संक्षिप्तमुक्तं पृथुताभयेन नान्तोऽस्ति यस्माद्गणितार्णवस्य॥ २७४॥

उदाहरणम्। पञ्चस्थानस्थितैरङ्कैर्यद्यद्योगे त्रयोदश। कति भेदा भवेत्सङ्ख्या यदि वेत्सि निगद्यताम्॥ २७५॥

अत्राङ्कैक्यम्। १३। निरेकम्। १२। एतन्निरेकस्थानान्तमेकापचितमेकादिभिश्च भक्तं जातम् १२/१ ११/२ १०/३ ९/४

एषां घातैः समा जाताः सङ्ख्याभेदाः। ४९५।

इति श्रीलीलावत्यामङ्कपाशः।

न गुणो न हरो न कृतिर्न घनः पृष्टस्तथापि दुष्टानाम्। गर्वित-गणकबहूनां स्यात्पातोऽवश्यमङ्कपाशेऽस्मिन् ॥ २७६ ॥

येषां सुजातिगुणवर्गविभूषिताङ्गी शुद्धाखिलव्यवहृतिः खलु कण्ठसक्ता। लीलावतीह सरसोक्तिमुदाहरन्ती तेषां सदैव सुखसम्पदुपैति वृद्धिम् ॥ २७७ ॥

[अष्टौ व्याकरणानि षट् च भिषजां व्याचष्ट ताः संहिताः षट् तर्कान् गणितानि पञ्च चतुरोवेदानधीते स्म यः। रत्नानां त्रितयं द्वयञ्च बुबुधे मीमांसयोरन्तरं सद्ब्रह्मैकमगाधबोधमहिमा सोऽस्याः कविर्भास्करः ॥]

इति श्रीभास्कराचार्य्यविरचिते सिद्धान्त-
शिरोमणौ लीलावतीनामकः
प्रथमः खण्डः सम्पूर्णः ॥